Math Activity Books

The Learner and the Learned Booklet

Thes Book Belongs To :

_____

_____

Maths Activity and Practice Book

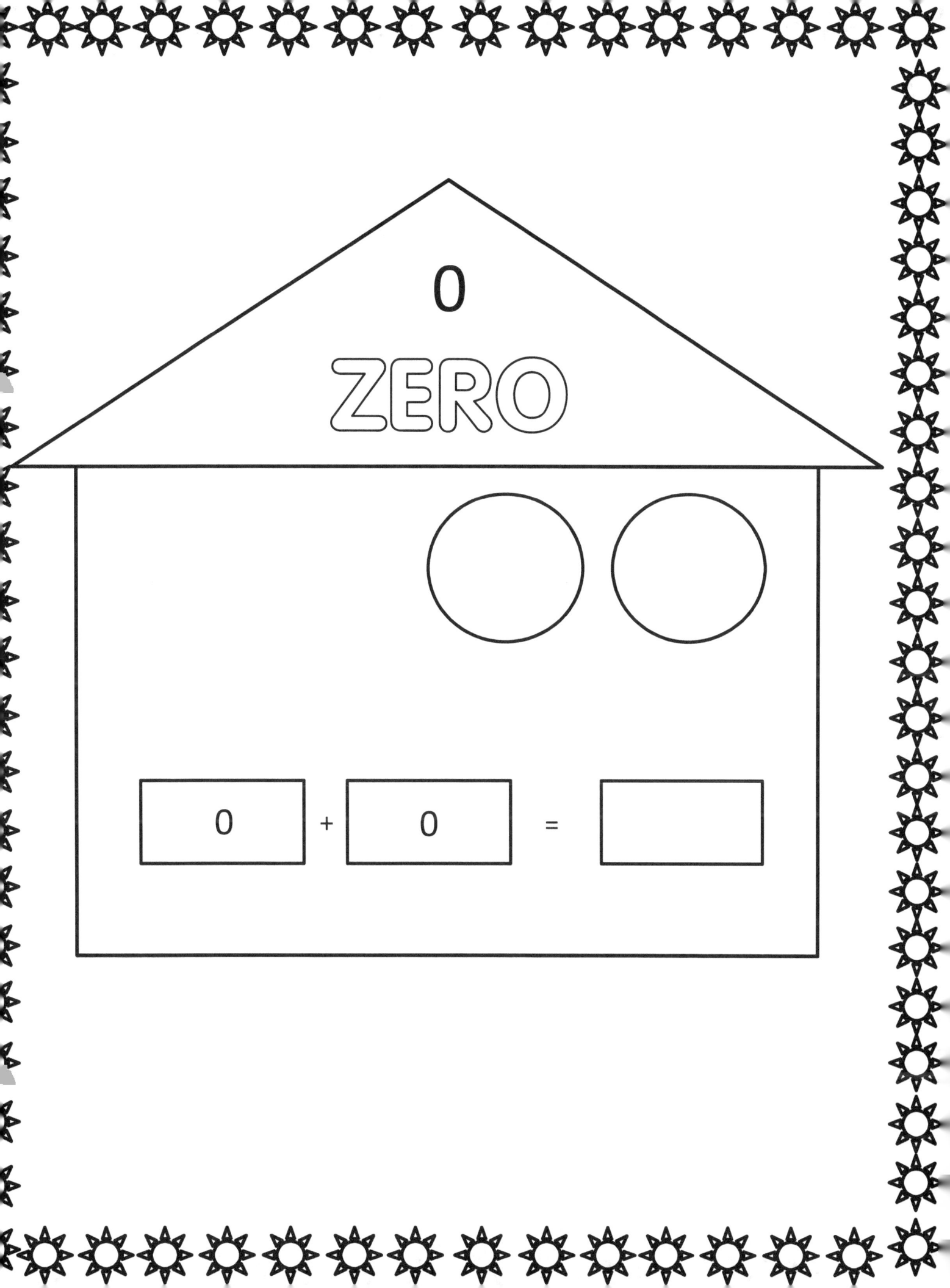

0

ZERO

0 + 0 =

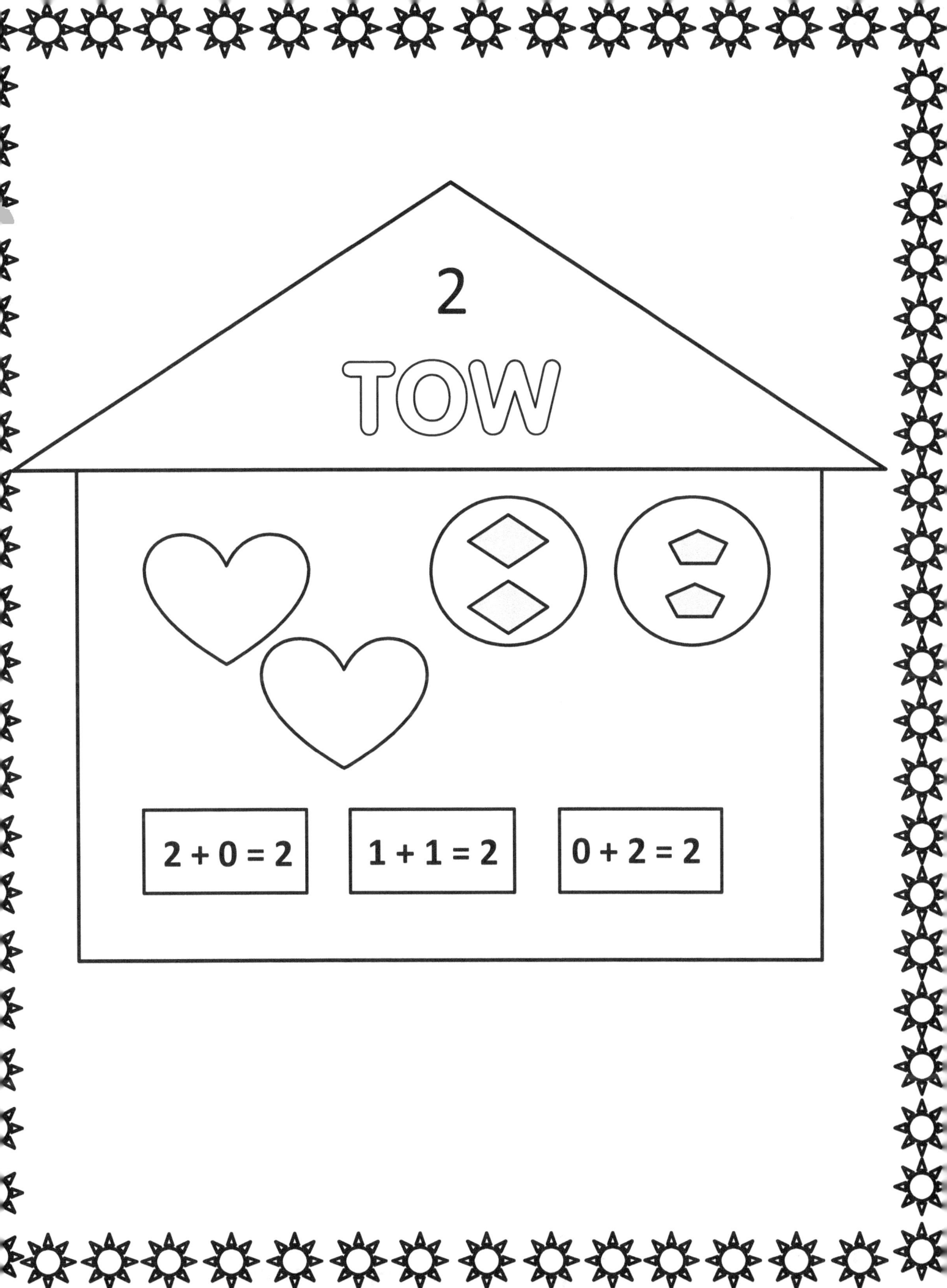

2

TOW

2 + 0 = 2   1 + 1 = 2   0 + 2 = 2

3

THREE

0 + 3 = 3

1 + 2 = 3

2 + 1 = 3

3 + 0 = 3

4

FOUR

4 + 0 = 4   0 + 4 = 4   2 + 2 = 4

3 + 1 = 4   3 + 1 = 4

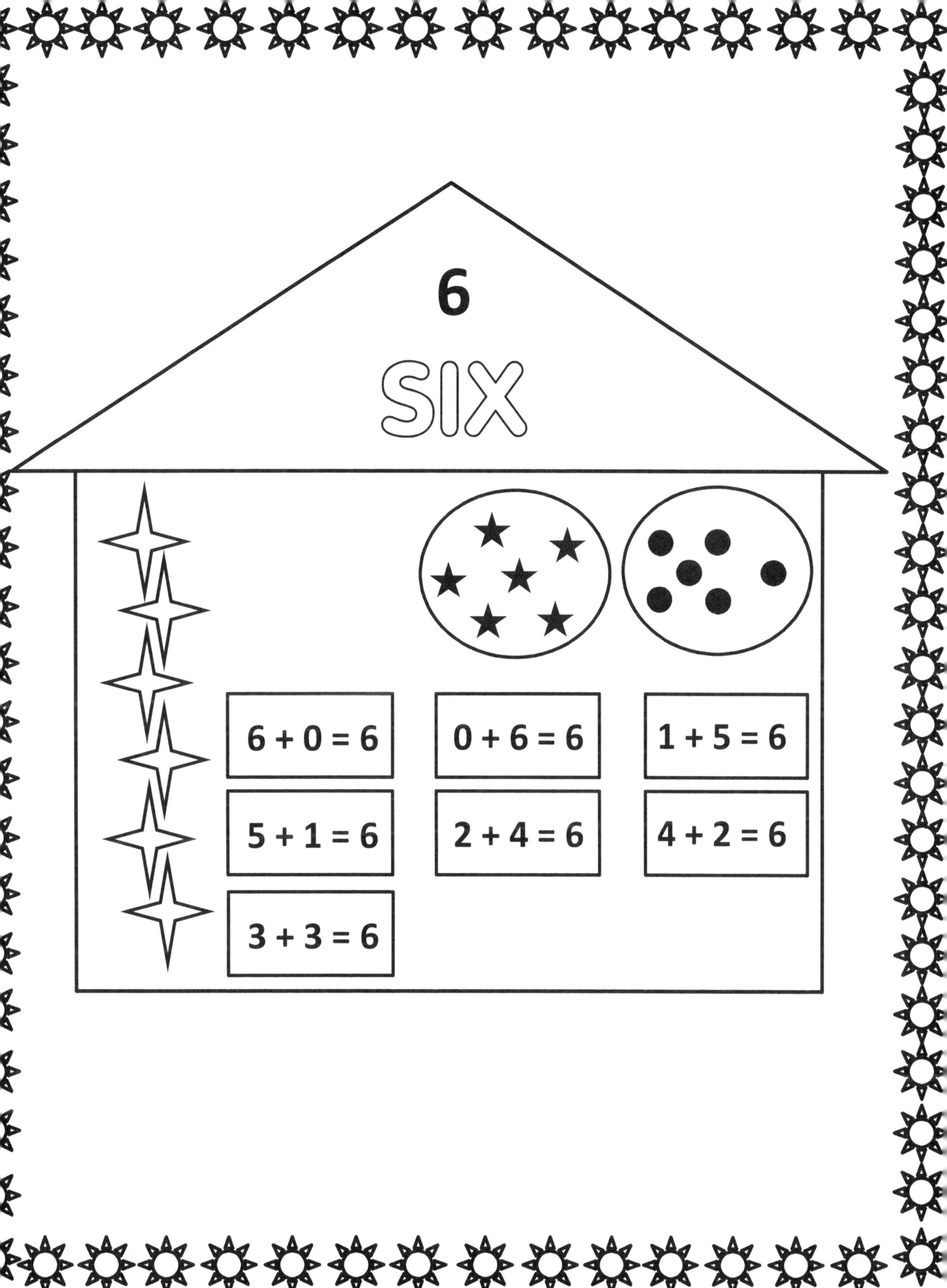

7

SEVNE

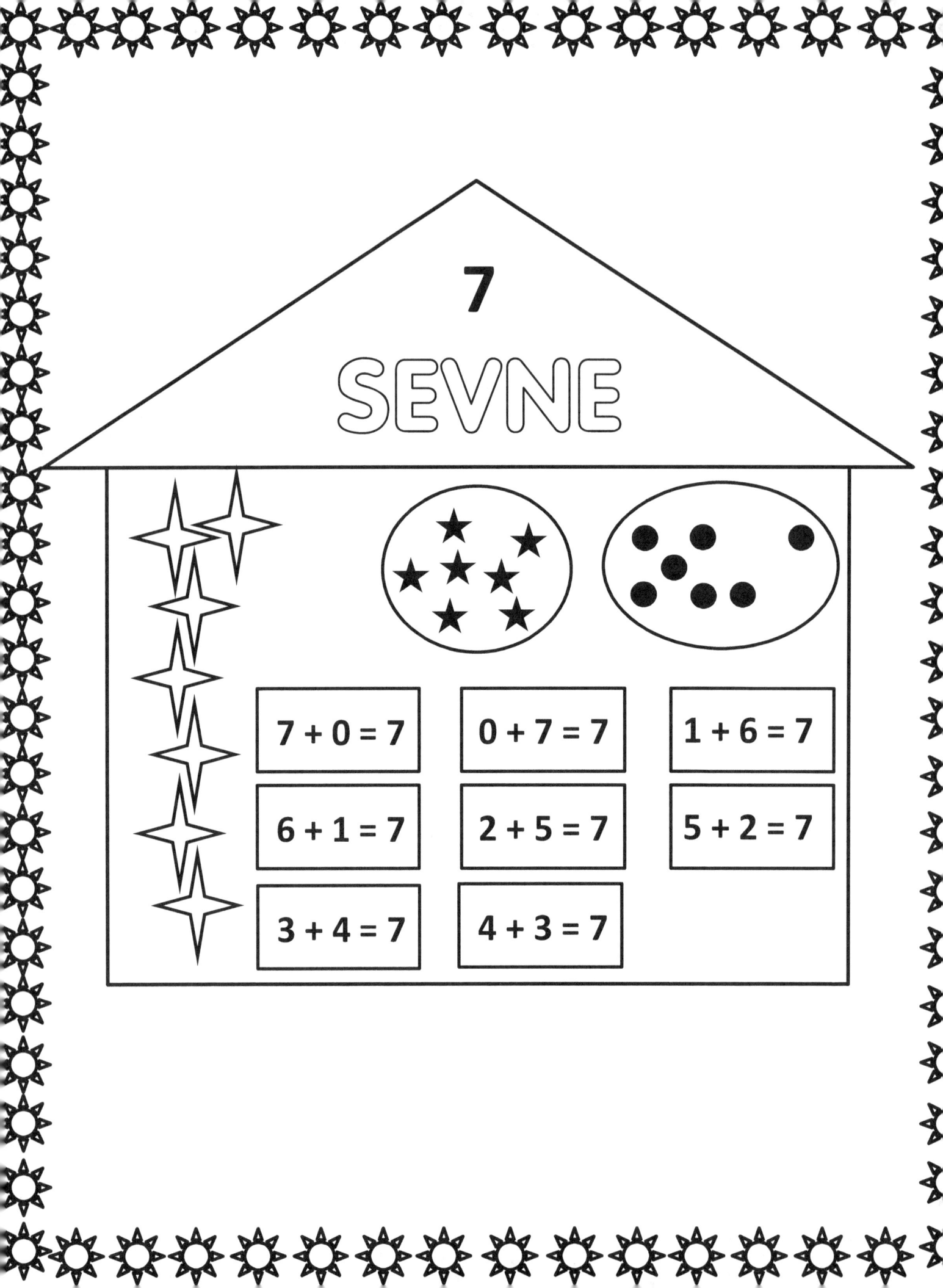

7 + 0 = 7	0 + 7 = 7	1 + 6 = 7
6 + 1 = 7	2 + 5 = 7	5 + 2 = 7
3 + 4 = 7	4 + 3 = 7	

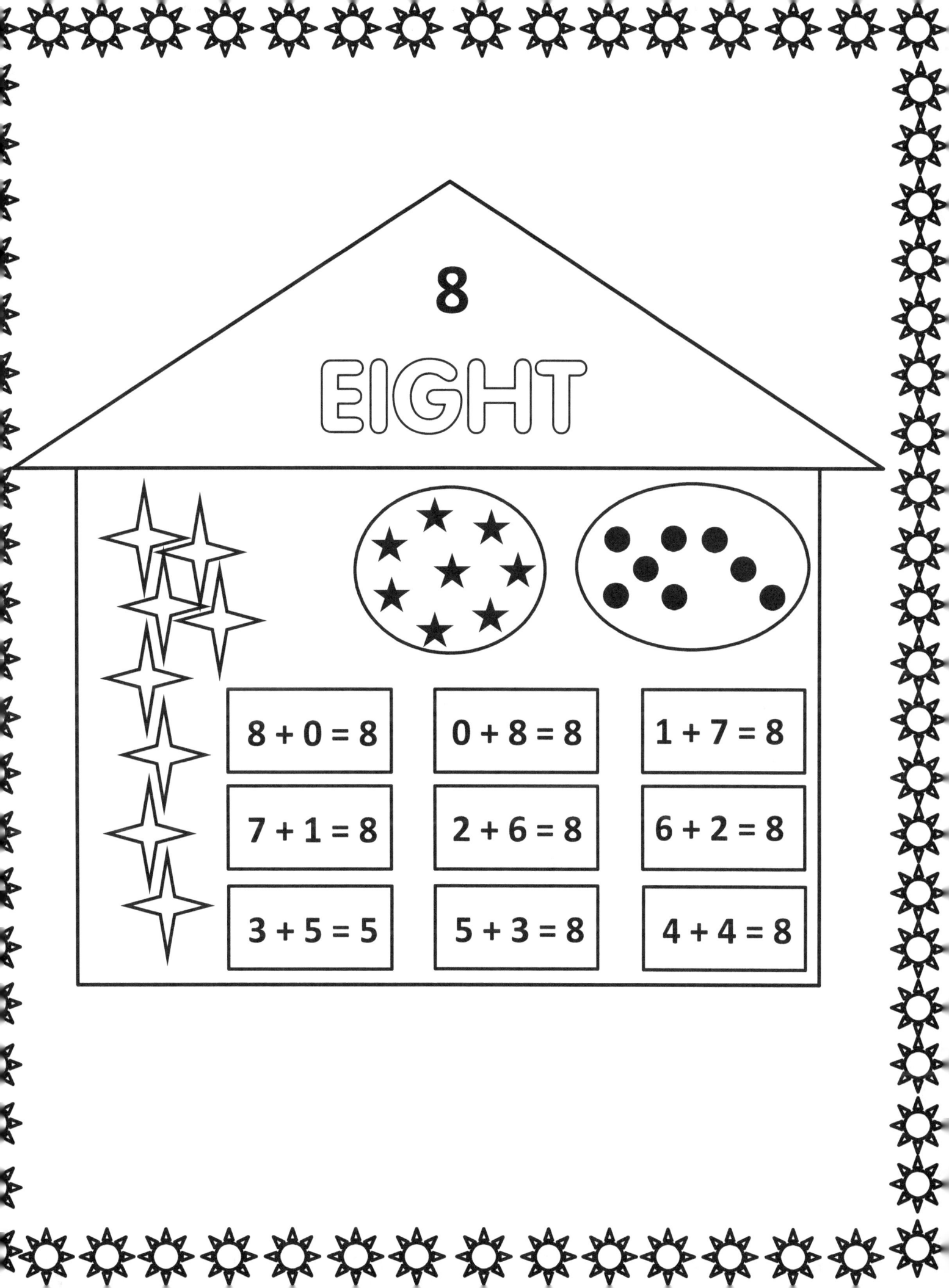

8

EIGHT

8 + 0 = 8	0 + 8 = 8	1 + 7 = 8
7 + 1 = 8	2 + 6 = 8	6 + 2 = 8
3 + 5 = 5	5 + 3 = 8	4 + 4 = 8

9

NINE

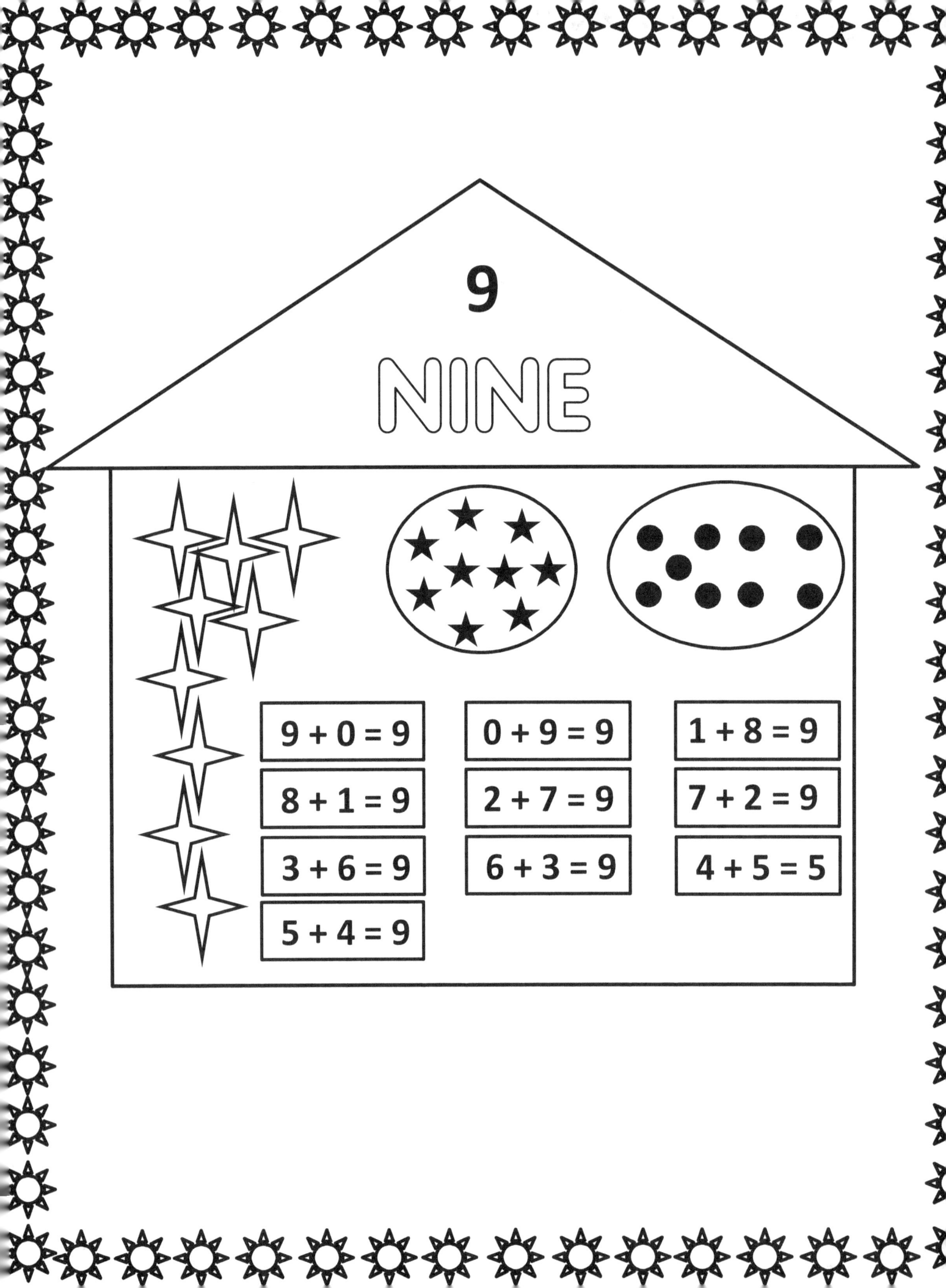

9 + 0 = 9	0 + 9 = 9	1 + 8 = 9
8 + 1 = 9	2 + 7 = 9	7 + 2 = 9
3 + 6 = 9	6 + 3 = 9	4 + 5 = 5
5 + 4 = 9		

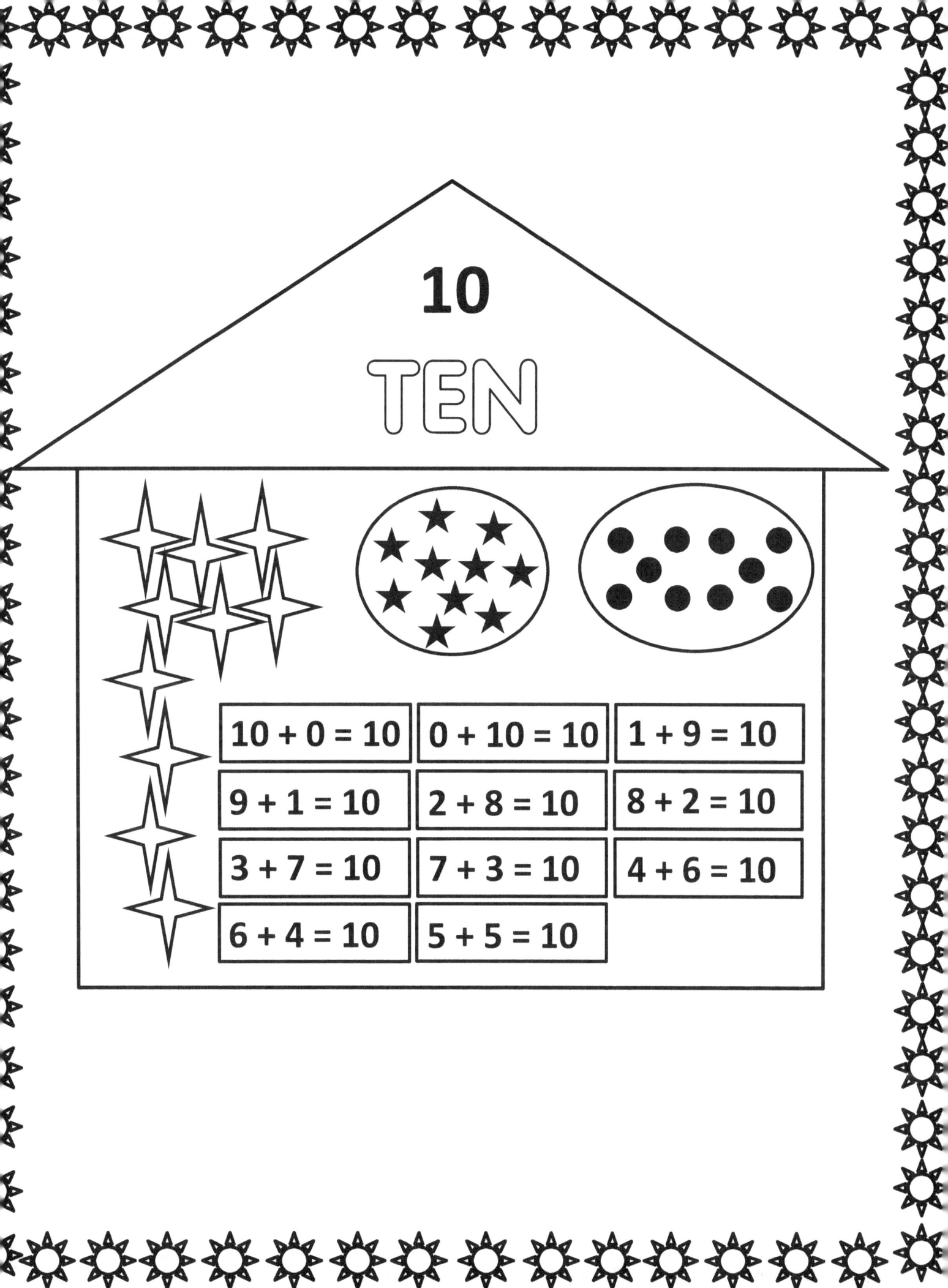

10

TEN

10 + 0 = 10	0 + 10 = 10	1 + 9 = 10
9 + 1 = 10	2 + 8 = 10	8 + 2 = 10
3 + 7 = 10	7 + 3 = 10	4 + 6 = 10
6 + 4 = 10	5 + 5 = 10	

Order and zero properties

Write the sum

4 + 7 = 11 7 + 4 = 11	3 + 5 = ... 5 + 3 = ...	0 + 1 = ... 1 + 0 = ...
9 + 2 = ... 2 + 9 = ...	8 + 6 = ... 6 + 8 = ...	4 + 2 = ... 2 + 4 = ...

4 +2 ___	2 +4 ___	1 +5 ___	5 +1 ___	6 +3 ___	3 +6 ___
7 +1 ___	1 +7 ___	0 +4 ___	4 +0 ___	8 +5 ___	5 +8 ___

▶ **Mixed review**

Solve

A. 5c + 2c = ...c

B. 8c + 3c = ...c

C. 8 + 2 = ...

D. 7 + 4 = ...

E. 3 + 1 = ...

F. 4 + 3 = ...

Count on 1 , 2 ,and 3

Circle the greater number
Count on to find the sum

a , $6 + 1 = \ldots\ldots$ b. $4 + 5 = \ldots\ldots$ c. $2 + 9 = \ldots\ldots$

d , $7 + 3 = \ldots\ldots$ e. $2 + 8 = \ldots\ldots$ f. $4 + 4 = \ldots\ldots$

4	4	4	4	4
+4	+4	+4	+4	+4

15	7	1	$+\ \ \begin{matrix}2\\4\end{matrix}$	4
+3	+5	+19		+4

7	1	6	5	12
+11	+8	+3	+4	+ 1

▶ **Mixed review**

Solve

A. $8c + 4c = \ldots\ldots c$ B. $4c + 8c = \ldots\ldots c$

C. $10 + 9 = \ldots\ldots$ D. $11 + 02 = \ldots\ldots$

Double And double plus one

 Vocabulary

Circle the **doubles plus one** fact in bleu
Circle the **doubles** fact in red

3 + 3 = 6   3 + 1 = 4   3 + 4 = 7

Complet the addition table :

+	0	1	2	3	4	5	6	7	8	9
0										
1										
2										
3										
4										
5										
6										
7										
8										
9										

Mak a Ten

Use a ten-forme and ◯ to make a ten.
Find the sum.

1

$+\ \dfrac{5}{5}$ $+\ \dfrac{6}{5}$ $+\ \dfrac{7}{7}$ $+\ \dfrac{8}{3}$ $+\ \dfrac{4}{7}$ $+\ \dfrac{6}{5}$

= . .   = . .   = . .   = . .   = . .   = . .

2

$+\ \dfrac{2}{9}$ $+\ \dfrac{9}{6}$ $+\ \dfrac{9}{9}$ $+\ \dfrac{4}{6}$ $+\ \dfrac{5}{8}$ $+\ \dfrac{6}{8}$

= . .   = . .   = . .   = . .   = . .   = . .

3

$+\ \dfrac{9}{5}$ $+\ \dfrac{7}{9}$ $+\ \dfrac{8}{4}$ $+\ \dfrac{7}{3}$ $+\ \dfrac{7}{5}$ $+\ \dfrac{8}{8}$

= . .   = . .   = . .   = . .   = . .   = . .

4

$+\ \dfrac{9}{3}$ $+\ \dfrac{2}{8}$ $+\ \dfrac{6}{6}$ $+\ \dfrac{4}{9}$ $+\ \dfrac{9}{1}$ $+\ \dfrac{7}{8}$

= . .   = . .   = . .   = . .   = . .   = . .

▶ **Mixed review**

Solve

5. $6 + 0 =$ . .   6. $0 + 10 =$ . .   7. $5 + 1 =$ . .

8. $3 + 4 =$ . .   9. $3 + 9 =$ . .   10. $2 + 4 =$ . .

11. $7 + 0 =$ . .   12. $4 + 4 =$ . .   13. $3 + 6 =$ . .

Add 3 Numbers

Circle the addends you add first , write the sun

1

6	6	7	8	4	6
5	5	7	3	7	5
+ 5	+ 2	+ 6	+ 3	+ 2	+ 3
= . .	= . .	= . .	= . .	= . .	= . .

2

2	9	9	4	5	6
9	6	9	6	8	8
+ 3	+ 1	+ 5	+ 5	+ 2	+ 4
= . .	= . .	= . .	= . .	= . .	= . .

3

9	7	8	7	7	8
5	9	4	3	5	8
+ 2	+ 4	+ 7	+ 1	+ 9	+ 2
= . .	= . .	= . .	= . .	= . .	= . .

4

9	2	6	4	9	7
3	8	6	9	1	8
+ 1	+ 0	+ 1	+ 3	+ 2	+ 5
= . .	= . .	= . .	= . .	= . .	= . .

▶ **Mixed review**

Solve

5. $5 + 2 = $ . .   6. $8 + 10 = $ . .   7. $8 + 7 = $ . .

8. $9 + 5 = $ . .   9. $5 + 1 = $ . .   10. $4 + 4 = $ . .

11. $4 + 6 = $ . .   12. $6 + 3 = $ . .   13. $5 + 7 = $ . .

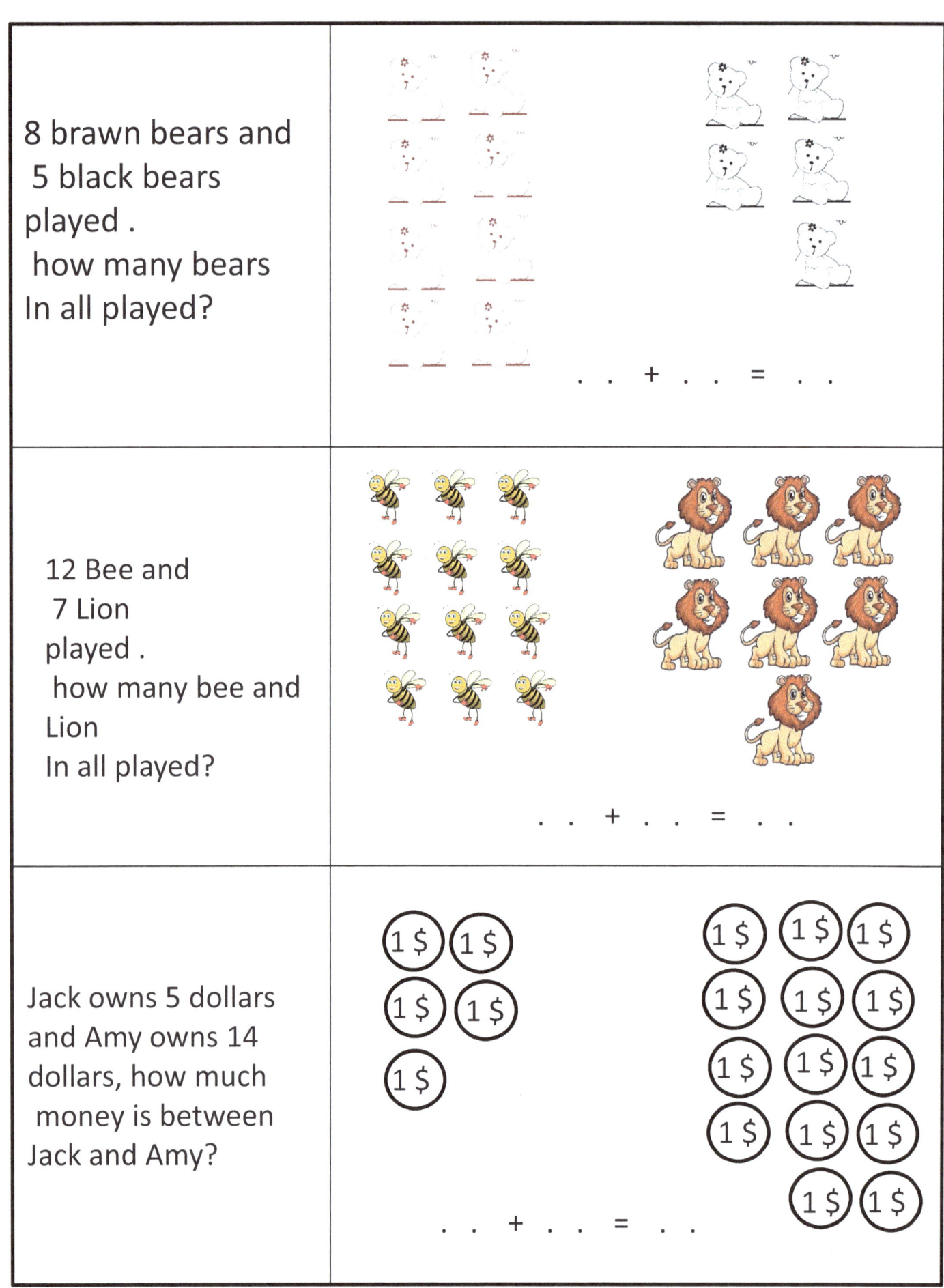

8 brawn bears and
 5 black bears
played .
 how many bears
In all played?

.. + .. = ..

12 Bee and
 7 Lion
played .
 how many bee and
Lion
 In all played?

.. + .. = ..

Jack owns 5 dollars
and Amy owns 14
dollars, how much
 money is between
Jack and Amy?

.. + .. = ..

Subtract all or zero

subtract

1. How many flowers are left ?	2. How many flowers are left ?
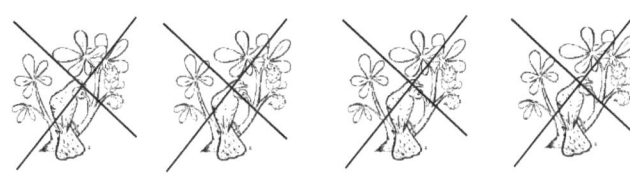	
4 − 4 = 0	4 − 0 = . .

3

− 5	− 4	− 7	− 5	+ 6	+ 1
0	4	0	5	0	1
= . .	= . .	= . .	= . .	= . .	= . .

4

− 8	− 8	− 2	− 2	− 3	− 6
8	0	2	0	3	6
= . .	= . .	= . .	= . .	= . .	= . .

5

− 1	− 7	− 3	− 9	− 9	− 4
0	7	0	9	0	0
= . .	= . .	= . .	= . .	= . .	= . .

▶ **Mixed review**

Solve

6. $7 + 2 =$ . .   7. $8 + 0 =$ . .   8. $3 + 6 =$ . .

9. $9 + 0 =$ . .   10. $5 + 1 =$ . .   11. $4 + 3 =$ . .

Count Back

Count back to find the difference

1. $7 - 2 = ..$ $8 + 2 = ..$ $6 - 4 = ..$

2. $6 - 3 = ..$ $9 - 3 = ..$ $5 - 1 = ..$

3.

11	10	13	12	20	15
- 05	- 07	- 08	- 02	- 04	- 06
= ..	= ..	= ..	= ..	= ..	= ..

4.

06	17	16	13	18	17
- 02	- 05	- 08	- 04	- 10	- 09
= ..	= ..	= ..	= ..	= ..	= ..

5.

14	11	15	11	06	16
- 03	- 09	- 07	- 11	- 05	- 04
= ..	= ..	= ..	= ..	= ..	= ..

6.   ?

11	12	14	70	30	17
- 10	- 14	- 05	- 10	- 05	- 15
= ..	= ..	= ..	= ..	= ..	= ..

▶ **Mixed review**

Solve

7. $70 + 20 = ..$ 8. $50 + 30 = ..$ 9. $30 - 15 = ..$

10. $90 - 40 = ..$ 11. $50 + 22 = ..$ 12. $55 - 5 = ..$

Number expressions

Look across down and diagonally,
Circle pairs of numbers that give the sum at the top,

16			
13	11	10	12
12	4	3	8
8	1	7	10
9	7	2	5
7	**4**	**11**	**1**

9			
6	3	5	0
1	2	4	8
2	5	1	2
3	6	7	1
4	**0**	**1**	**8**

Circle pairs of numbers that give the difference at the top

7			
7	8	5	12
0	4	6	15
3	1	8	4
8	1	9	2
4	**12**	**11**	**4**

6			
1	3	7	12
4	3	6	6
11	8	9	3
5	0	8	1
10	**9**	**2**	**15**

Remember Addition Facts

Write the sum

Use doubles	
5	10
3	
9	
8	

Use doubles Plus one	
6	
7	
4	

Add 0	
4	
6	
3	
2	

Count on 3	
8	
5	
7	
6	

Count on 2	
9	
3	
10	
6	

Count on 1	
8	
9	
4	
2	

▶ **Mixed review**

Solve

7. $20 + 10 + 1 = \ldots$

8. $1 + 13 + 7 = \ldots$

9. $14 - 12 - 2 = \ldots$

10. $90 - 40 - 10 = \ldots$

11. $23 + 22 = \ldots$

12. $55 - 5 - 10 = \ldots$

13. $12 - 5 - 3 = \ldots$

14. $80 - 27 = \ldots$

15. $25 - 5 - 15 = \ldots$

Missing Numbers

write the missing number.
use counters if you need to.

1. $7 - .. = 4$ $8 + .. = 12$ $6 - .. = 2$

2. $.. - 3 = 4$ $.. - 3 = 3$ $.. - 1 = 8$

3. $9 - .. = 5$ $2 + .. = 10$ $5 - .. = 5$

4. $.. - 5 = 8$ $.. - 4 = 7$ $.. - 2 = 9$

5. $6 - .. = 4$ $20 + .. = 35$ $12 - .. = 2$

6. $.. - 13 = 2$ $.. - 3 = 20$ $.. - 4 = 9$

7. $5 - .. = 1$ $12 + .. = 30$ $9 - .. = 3$

8. $.. - 12 = 5$ $.. - 4 = 0$ $.. - 7 = 1$

9. $7 - .. = 7$ $8 + .. = 9$ $11 - .. = 4$

10. $.. - 15 = 10$ $.. - 5 = 17$ $.. - 2 = 7$

Numbers 0 to 99

1. Write each number on the appropriate card 59, 32, 16, 37, 25, 43

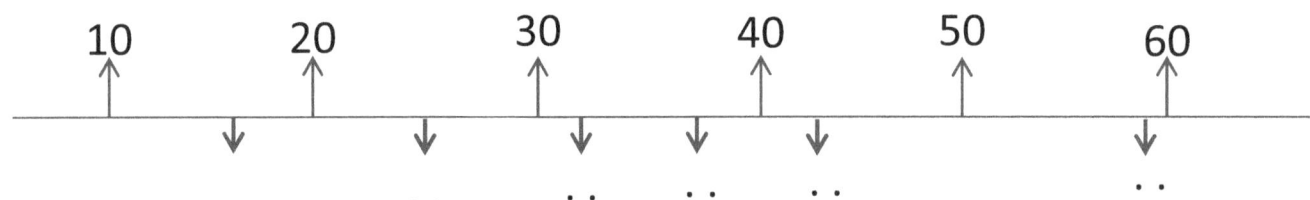

2. Note and complete: The number 25 is sandwiched between the two numbers
..... and.....

The number 59 is sandwiched between the two numbers
..... and.....

Write in the numbers in the table the number represented:

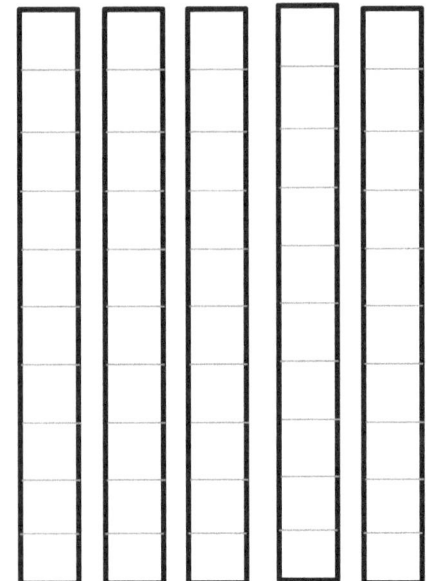

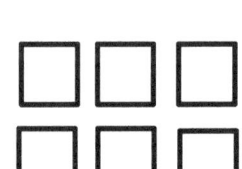

Tens	Units
. .	. .

Write in numbers:

twenty-four : [　　　]

thirty-one : [　　　]

seventeen : [　　　]

Sixteen : [　　　]

write in letters :

88 : [　　　]

42 : [　　　]

30 : [　　　]

19 : [　　　]

Read and fill in the table:

	T	U
Twenty-six		
Forty-three		
Eighty-four		

Mental Arithmetic

1. 40 - 13 = . .   2. 89 + 11 = . .   3. 18 - 05 = . .

I notice and write

Right Before Him	The Number	Right After It
	52	
39		
		92
	20	

Put the symbol (< or >) in the appropriate place:

12 ,,,,,, 15   ;   18 ,,,,,, 25   ;   96 ,,,,,, 14   ;   12 ,,,,,, 21

82 ,,,,,, 24   ;   85 ,,,,,, 72   ;   02 ,,,,,, 20   ;   30 ,,,,,, 20

01 ,,,,,, 10   ;   52 ,,,,,, 48   ;   35 ,,,,,, 71   ;   90 ,,,,,, 24

25 ,,,,,, 26   ;   62 ,,,,,, 61   ;   47 ,,,,,, 74   ;   13 ,,,,,, 91

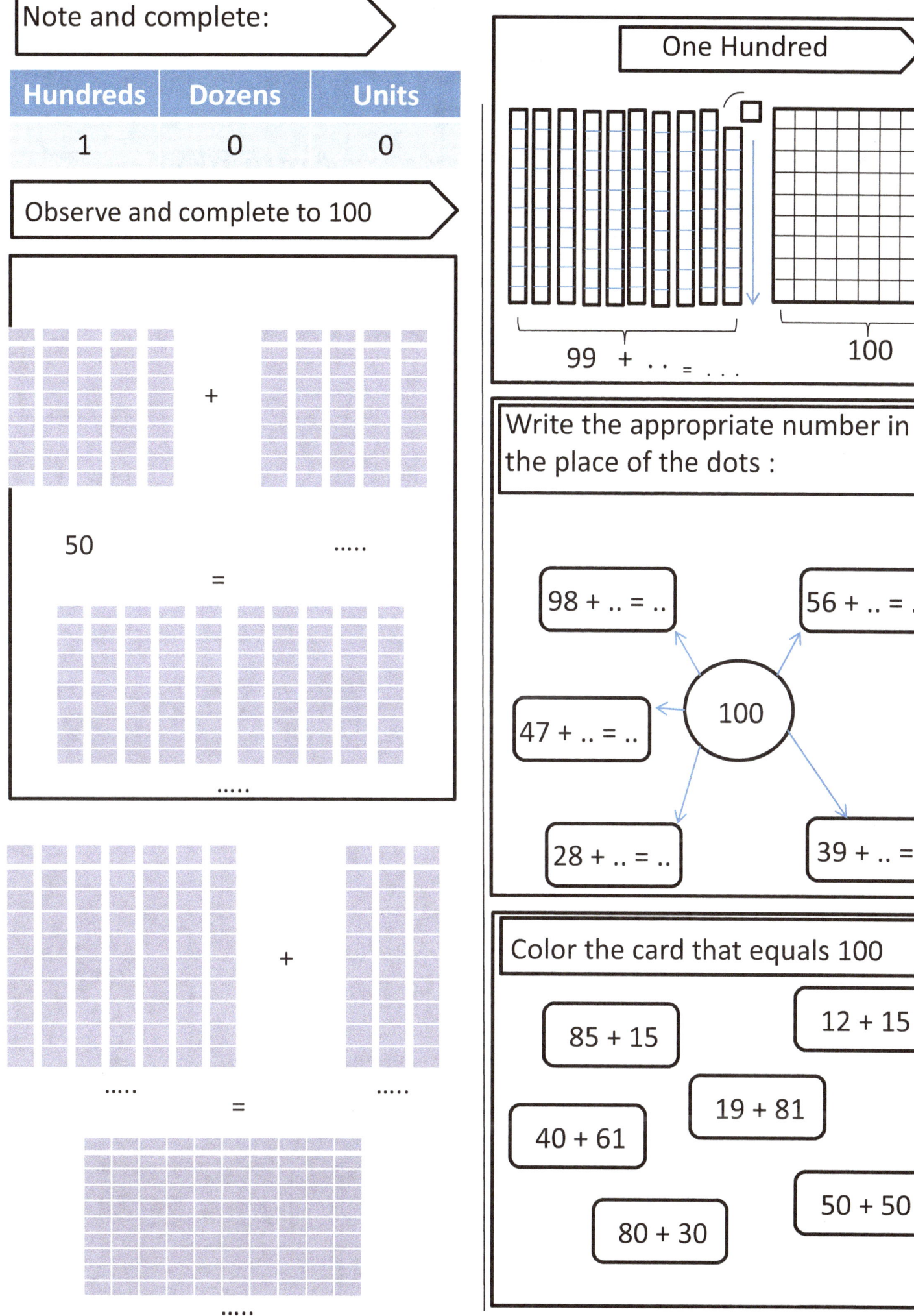

Note and complete:

Hundreds	Dozens	Units
1	0	0

Observe and complete to 100

+

50   .....

=

.....

+

....

=

.....

One Hundred

99 + . . = . . .   100

Write the appropriate number in the place of the dots :

98 + .. = ..   56 + .. = ..

47 + .. = ..   100   39 + .. = ..

28 + .. = ..

Color the card that equals 100

85 + 15   12 + 15

19 + 81

40 + 61

50 + 50

80 + 30

20 $ **50 $**

20 $ **1 $**

Amount is: ....

..... + ... + ... + ... = ...

Observe and write :

10		30						90	

100				60					10

Color the number 100 in red :

010

001

95 + 06

50 + 50

01 + 09

60 + 39

100

90 + 09 + 01

71 + 29

12 + 19 + 67

Create Geometric Shapes:

Observe the drawn shapes and write the numbers of the shapes that have:

+ Three sides:

....

+ Her name: ....

+ Four equal sides:

....

+ Her name: ....

+ The rectangle numbers are:

....

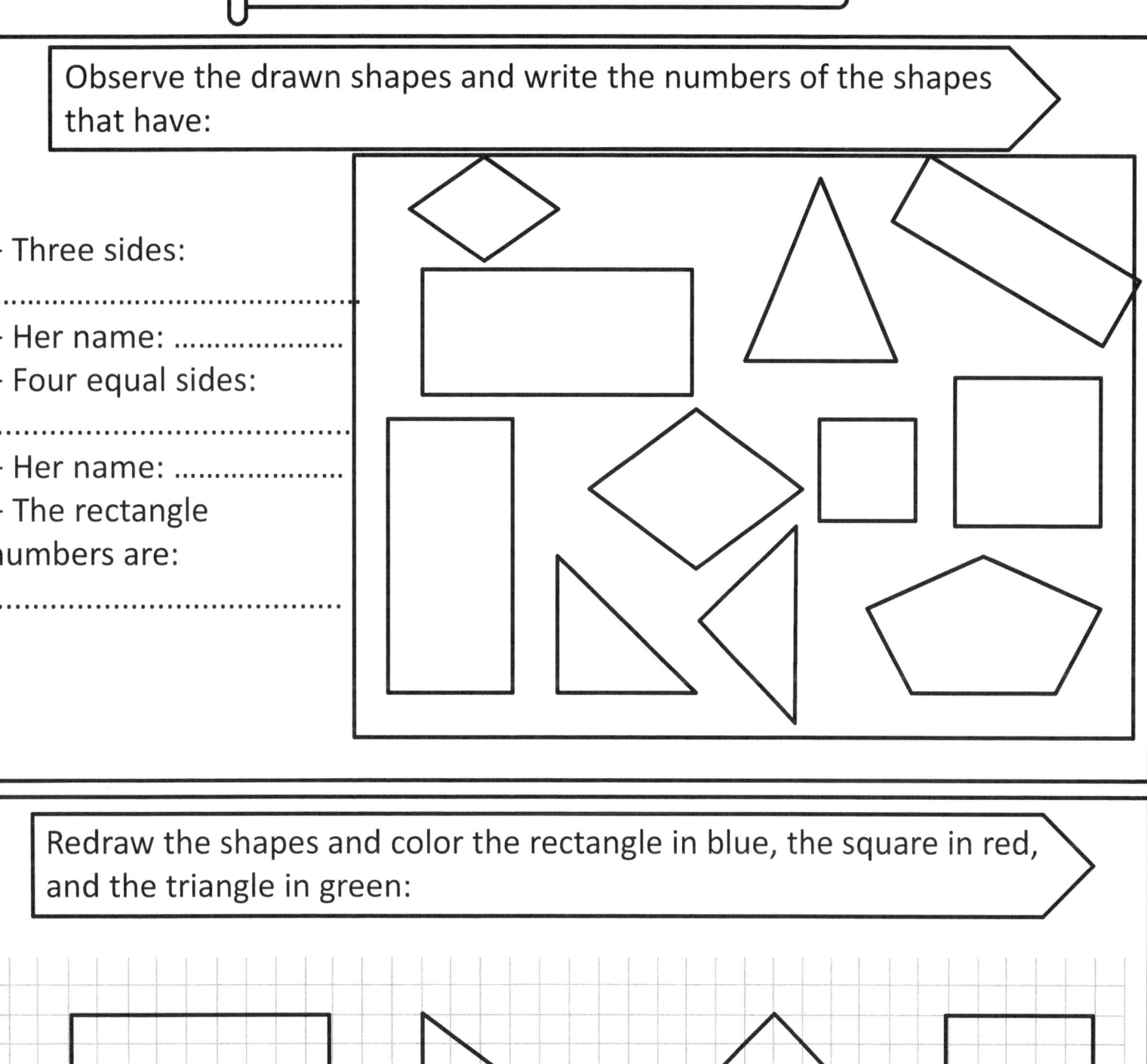

Redraw the shapes and color the rectangle in blue, the square in red, and the triangle in green:

Draw 3 squares, 5 triangles, and 4 rectangles:
Color the geometric shapes in the colors you love.

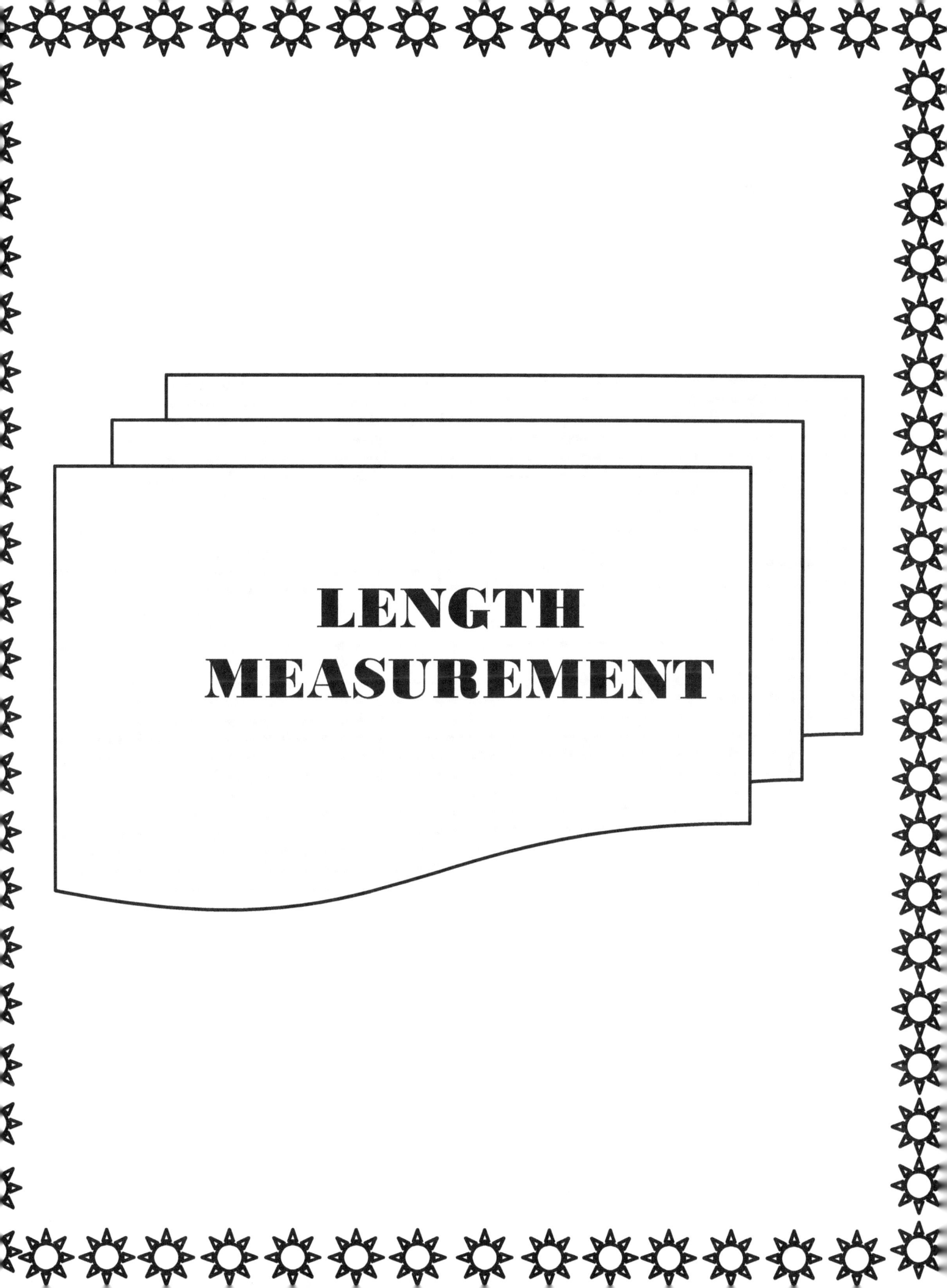

LENGTH MEASUREMENT

1 - The teacher asked Aslal and Amir to measure the length of the same pen.
Take note and complete.

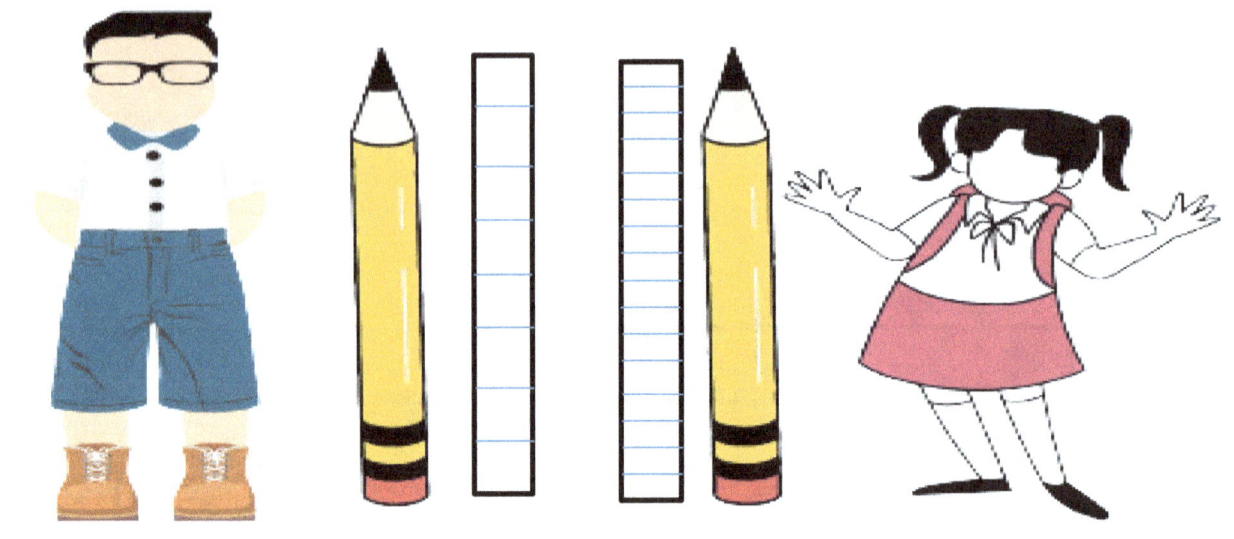

| The measurement of the length of the pen is: . . . . □ | The measurement of the length of the pen is: . . . . ▭ |

2 - Note and complete:

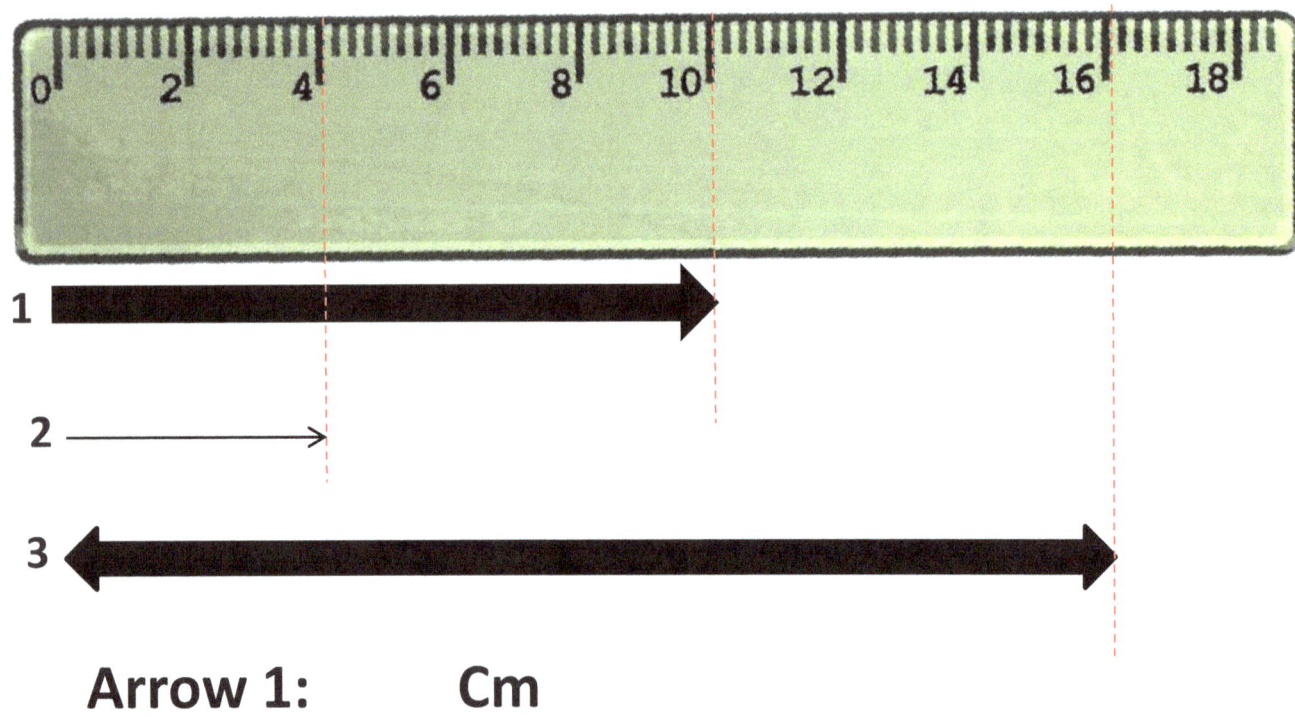

Arrow 1:   Cm

Arrow 2:   Cm

Arrow 3:   cm

3 - Observe and complete:

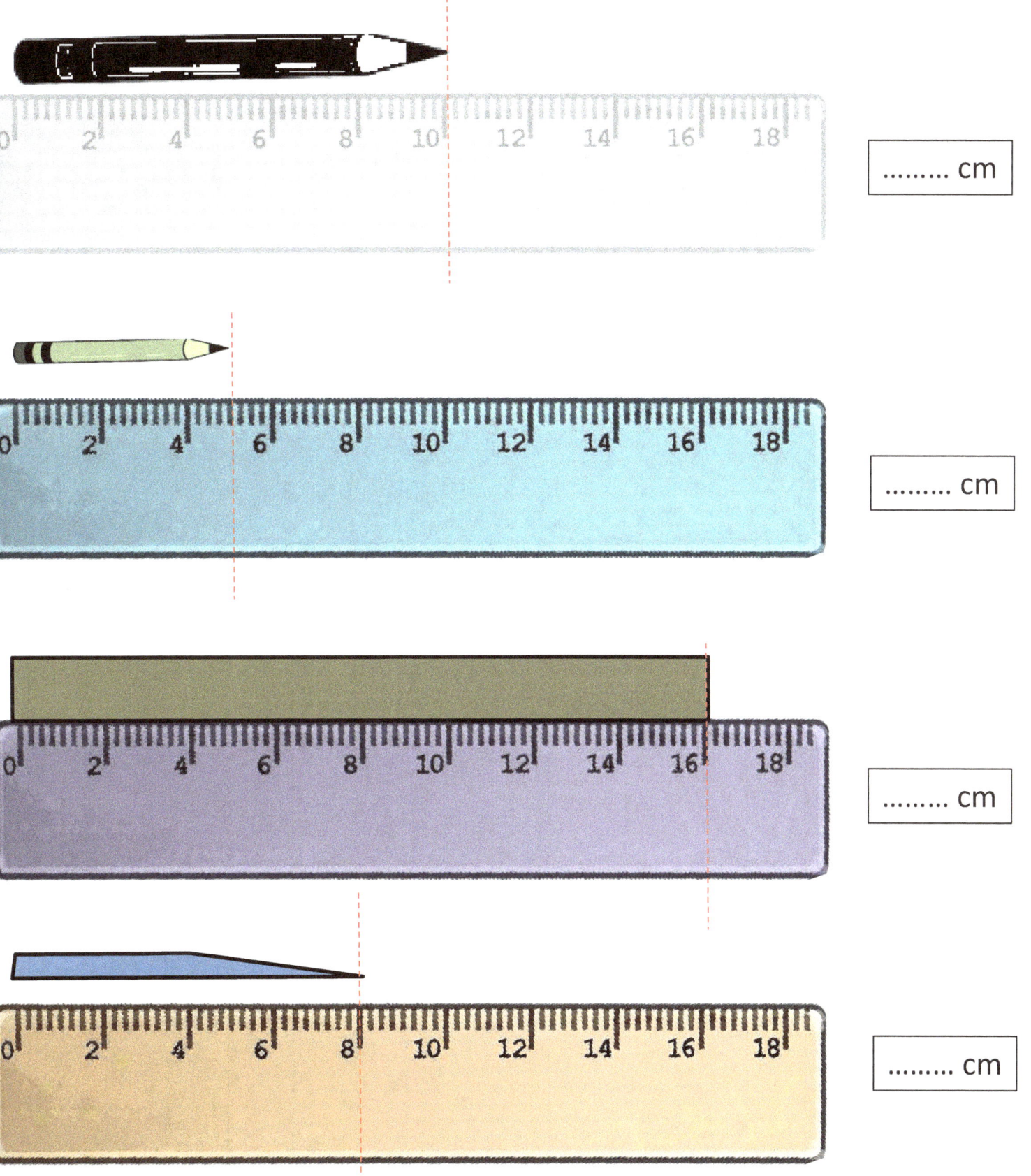

.... cm

.... cm

.... cm

.... cm

ORDINARY TECHNIQUE OF COLLECTING

Calculate 44 + 25 using the table:

D	U
....	....
....	....
....	....

Portrait mode :

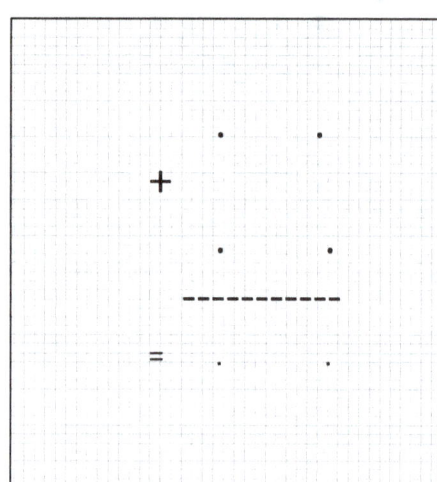

Put and do:

42 + 13

36 + 25

57 + 16

71 + 23

63 + 25

47 + 29

39 + 08

90 + 10

Surround in blue the largest number and in red the smallest number

120	502	914
195	209	854

The numbers between 311 and 603 are outlined with a black line

603	701	125
502	628	311

Put the appropriate symbol < or > in the appropriate place:

85 954	452 698	102 91	72 524

Write the numbers in incremental order on the bar (from smallest to largest):

200 ; 450 ; 320 ; 705 ; 12 ; 807 ; 633 ; 185 ; 528

....	....	....	365	....	....	....	....	....	....

Write in each star the appropriate number of the following: 338 – 447 – 293 - 402

270   320   370   420   470

Calculate mentally:

100 + 10 / 100 + 20 / 100 + 32 / 100 + 36 / 100 + 54 / 100 + 80 /
100 + 60 / 100 + 58 / 100 + 14 / 100 + 87 / 100 + 81 / 100 + 50 .

Calculate:

```
  1 0 2          4 5 7          2 5 6          3 4 1
+                +              +              +
      4 1            1 8 5          1 2 2          2 7 1
-------------   -------------   -------------   -------------

=               =              =              =
```

```
  8 0 5          7 6 3          5 2 1          2 2 2
+                +              +              +
  0 7 9            1 5 2          4 0 0          2 6 3
-------------   -------------   -------------   -------------

=               =              =              =
```

```
  5 1 4          2 4 1          3 4 2          7 1 9
+                +              +              +
  1 8 4            2 1 1          2 9 9          2 4 5
-------------   -------------   -------------   -------------

=               =              =              =
```

```
  3 6 4          2 8 2          6 2 4          6 2 4
+                +              +              +
  1 0 8            0 4 9          3 0 7          3 2 1
-------------   -------------   -------------   -------------

=               =              =              =
```

Put and do:

412 + 123

212 + 23

357 + 159

258 + 153

751 + 89

651 + 342

512 + 300

703 + 214

410 + 21

258 + 48

300 + 299

612 + 350

384 + 241

741 + 58

199 + 87

740 + 67

Camara owns $250.
I observe the photo and figure out what Sarah has.

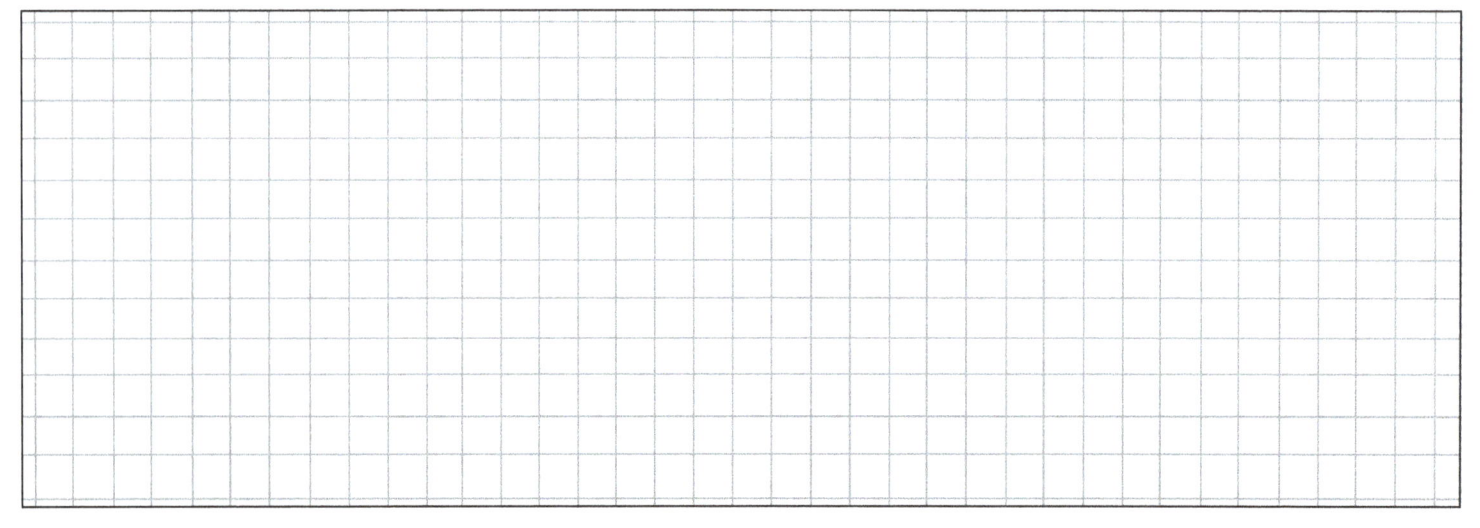

Lala has 36 blue, 147 red and 400 yellow agates.

Calculate the number of agate Lala.

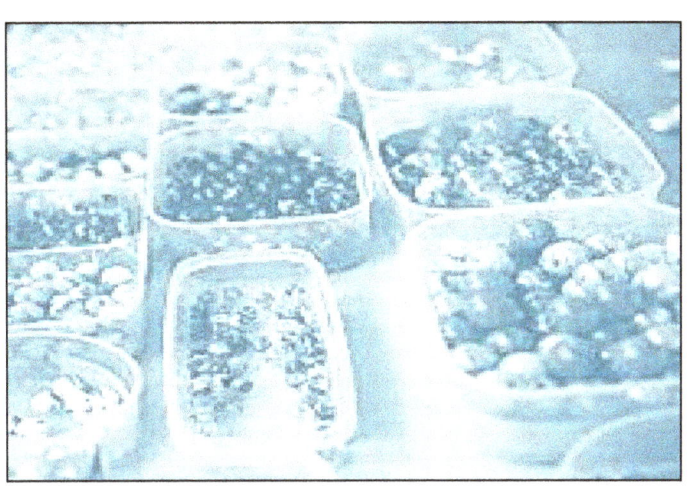

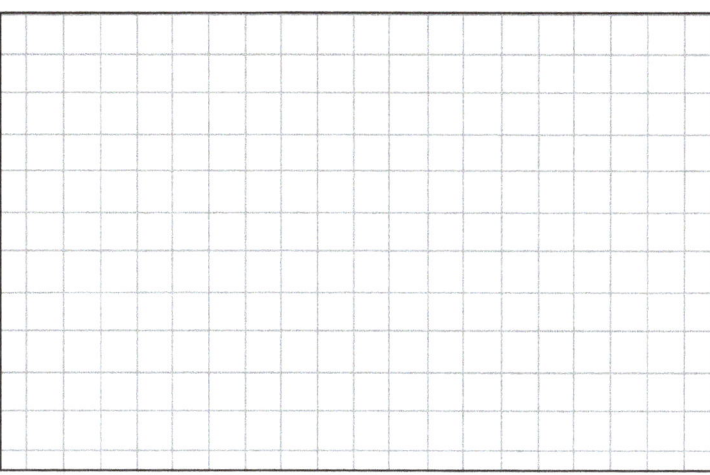

Calculate:

```
    2 , 4
+
    2 3 5
---------------
    , 5 ,
=
```

```
    1 7 7
+
    , 8 ,
---------------
    3 6 0
=
```

```
    2 , ,
+
    , 4 3
---------------
    5 7 6
=
```

```
    5 4 4
+
    , , ,
---------------
    9 5 7
=
```

Put and do:

68 + 23

33 + 25

58 + 24

79 + 12

Standard subtraction technique

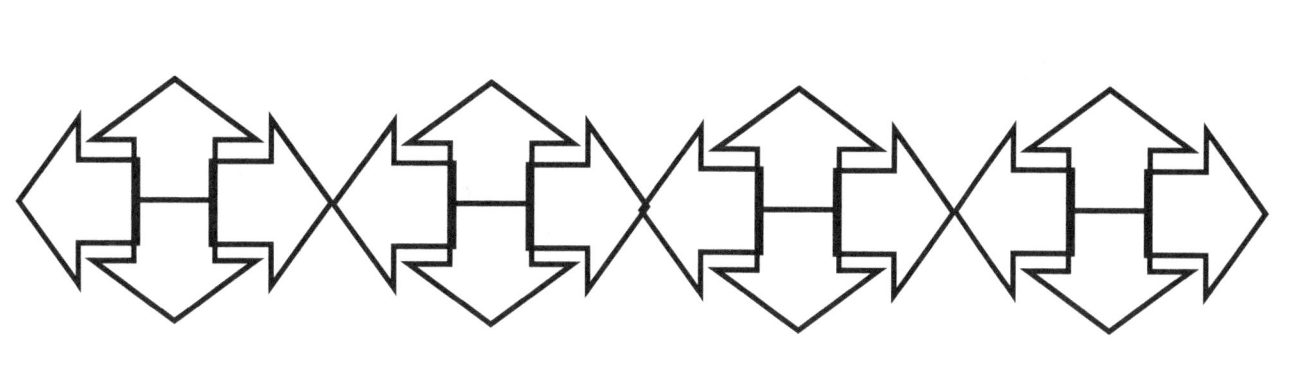

The world of numbers is extremely vast, and that is the main idea students explore in Grade 2. They extend their knowledge of numbers from 100 To 1000 ans review opérations they learned in Grade 1 with larger numbers. The fluency in addition and soustraction students attaining this year creates a firm fondation for the multiplication and division they will encounter in Grande 3.

Cross Out And Calculate:

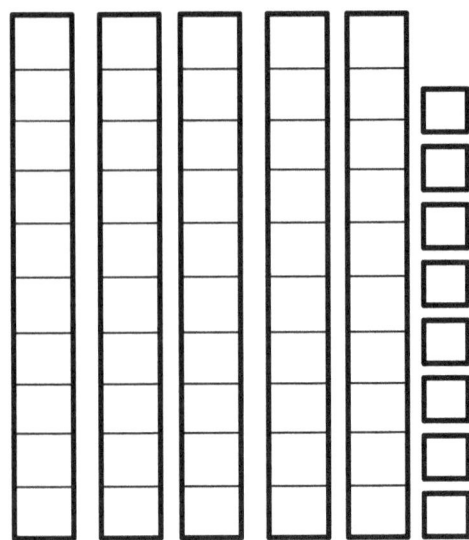

58 - 22 =

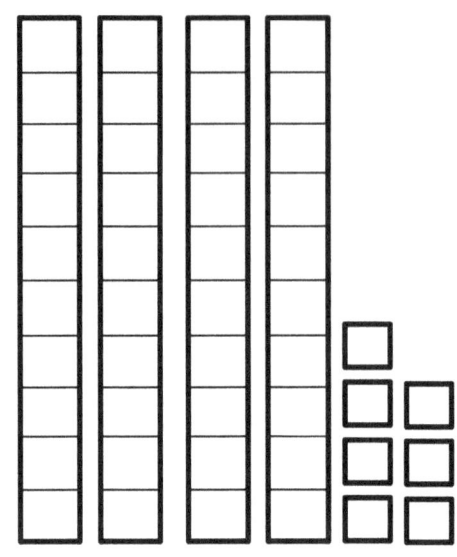

47 - 32 =

Accomplish:

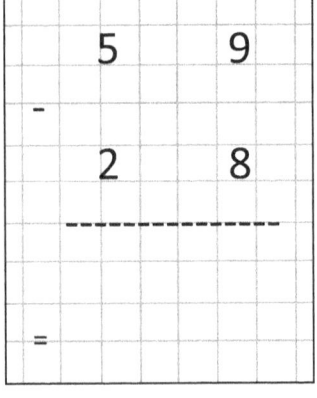

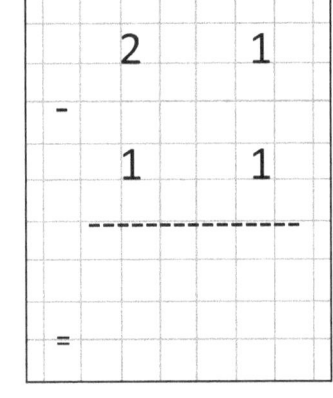

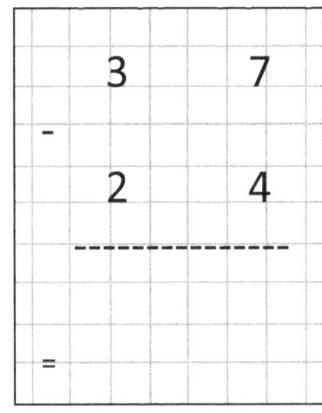

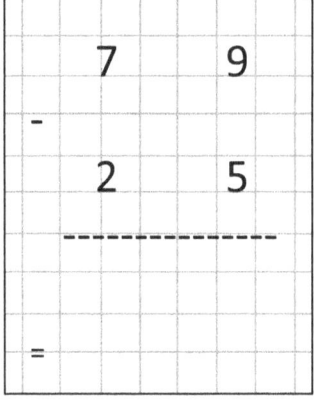

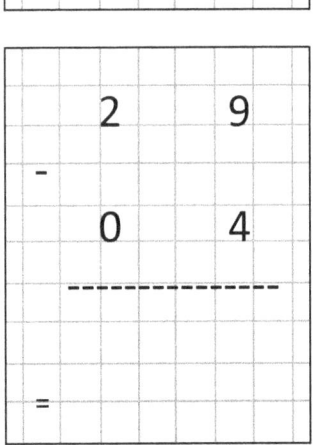

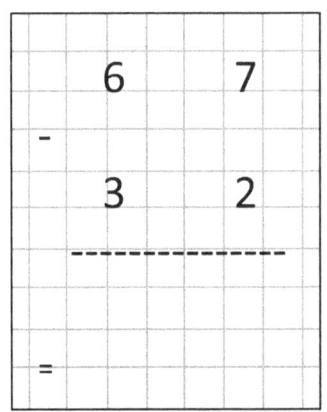

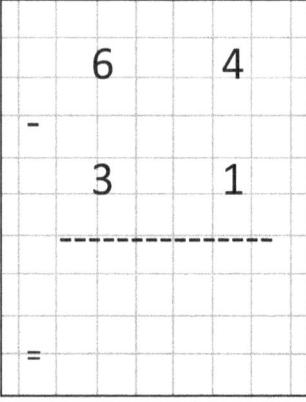

Calculate:

```
   3 8 1          1 2 4          4 7 0          1 2 1
 -   4 5        -   5 1        - 2 5 1        -   4 3
 -----------    -----------    -----------    -----------

 =              =              =              =
```

```
   9 7            6 3            9 1 2          1 8 6
 - 2 4          - 5 4          - 4 1 4        - 2 7 2
 -----------    -----------    -----------    -----------

 =              =              =              =
```

```
   4 5 1          8 0 1          6 2 1          4 5 1
 - 3 4 8        - 4 2 9        - 5 4 4        - 0 3 8
 -----------    -----------    -----------    -----------

 =              =              =              =
```

```
   6 3 4          7 8 9            2 9            4 5 4
 - 2 3 8        - 2 3 9        -   1 7        -   2 9
 -----------    -----------    -----------    -----------

 =              =              =              =
```

Put and do:

914 - 254

425 - 56

412 - 201

285 - 75

720 - 465

151 - 452

562 - 250

442 - 214

630 - 95

628 - 41

261 - 69

360 + 210

624 - 141

341 - 88

622 - 527

743 - 62

Write the appropriate number in the place of each dot :

	2	.
+		
	2	4
	,	,
=		

	.	9
+		
	4	.
	2	1
=		

	6	.
+		
	5	2
	,	3
=		

	7	.
+		
	3	4
	,	,
=		

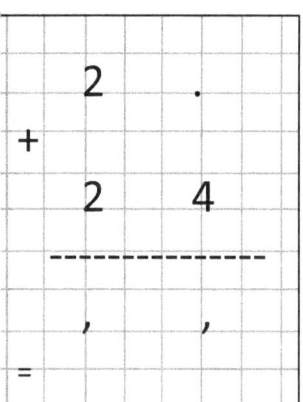

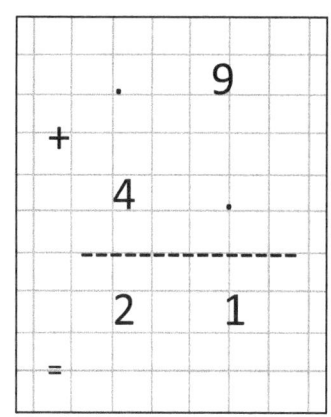

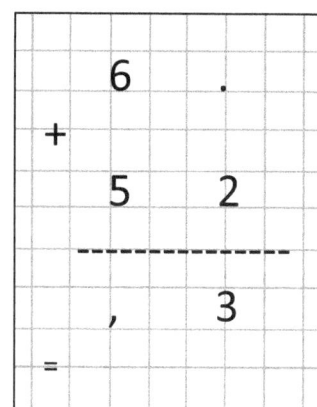

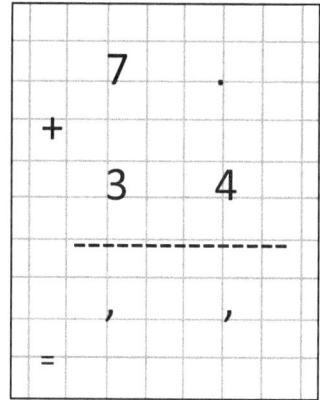

Put and do:

32 - 12 / 84 - 14 / 28 - 15 / 24 - 12 / 95 - 23 / 91 - 10

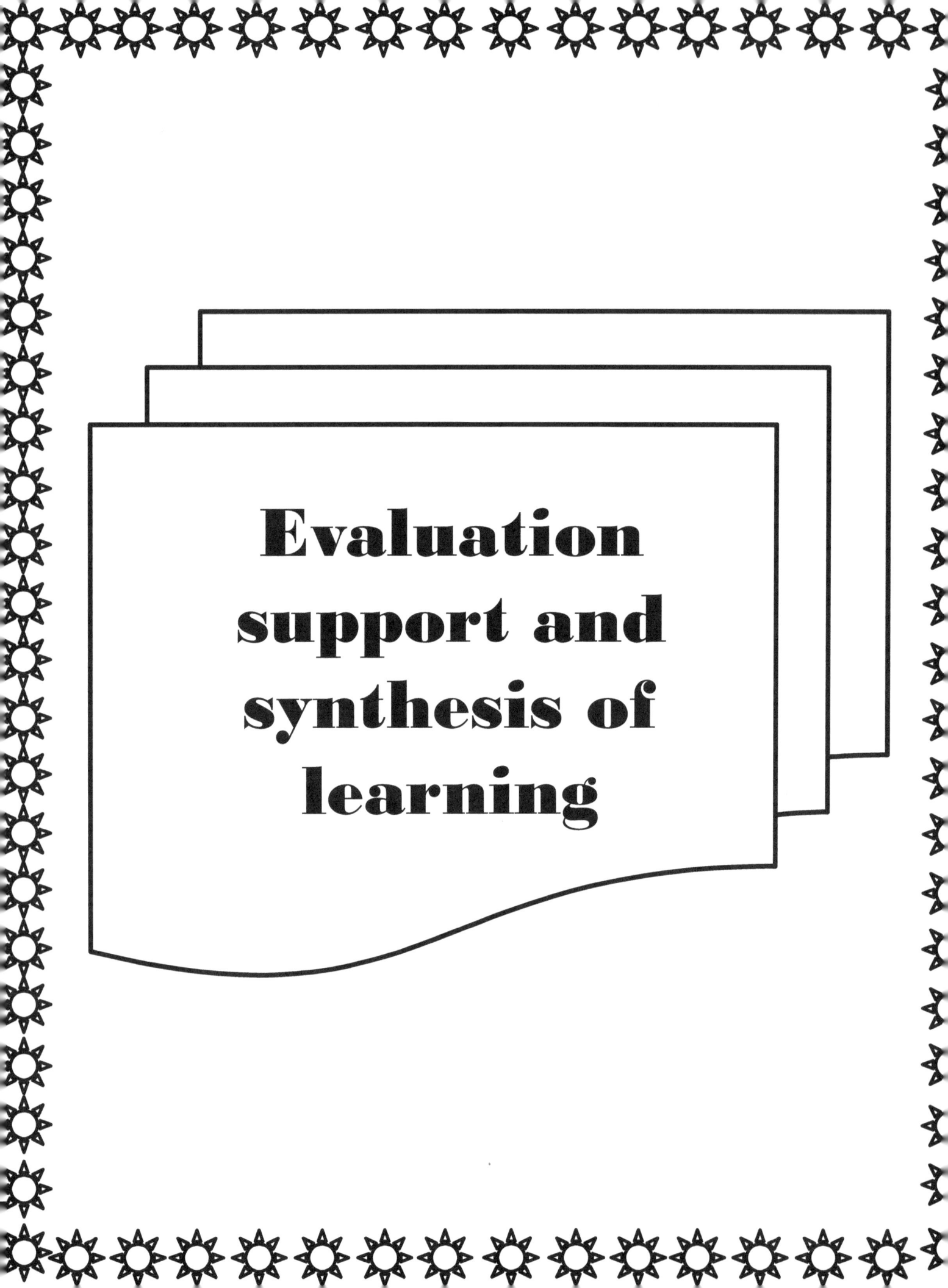

Evaluation support and synthesis of learning

1 - Color in green the card of the largest number and in yellow the card of the smallest number

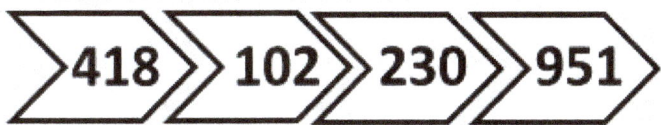

418 >> 102 >> 230 >> 951

2 - Then arrange these numbers in incremental order:

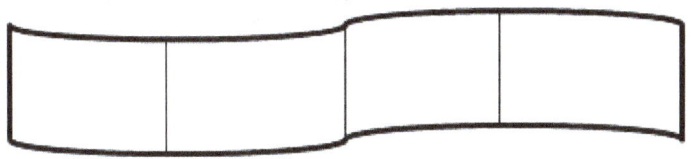

3 - Complete :

7 1 6	8 6	3 7
+	+	+
, ,	, ,	, ,
---------	---------	---------
7 5 3	9 8	1 6

4 - cross out the wrong writing

432	432

5 - Do addition and subtraction

5 6	1 4 1	3 2
-	+	+
2 3	3 9	1 5
_____	_____	_____
. .	. . .	. .

6 - Correct the situation and then perform the operation

8 9	1 2 6	2 6
-	+	+
4 3	5 9	2 4 9
_____	_____	_____

7 - Write on the card the appropriate measure of the red piece:

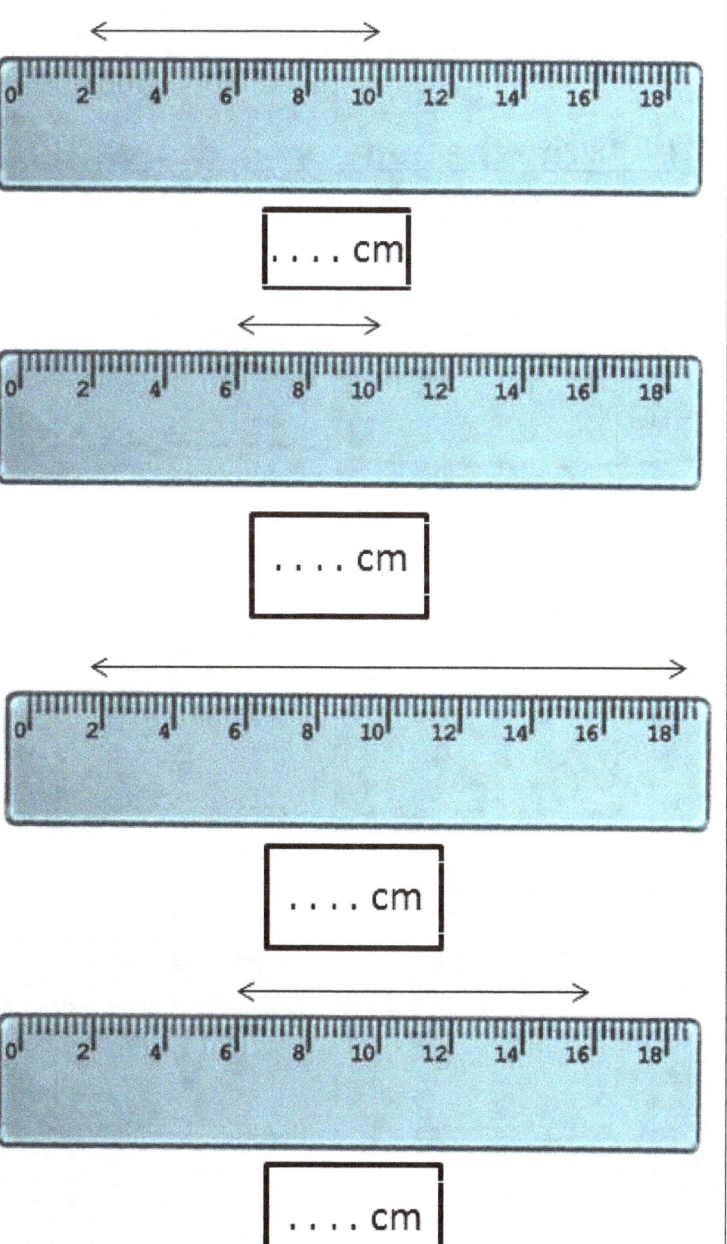

. . . . cm

. . . . cm

. . . . cm

. . . . cm

8 - Compare using one of the two symbols: < or >

512	342		624	410

9 - Complete by writing the appropriate number in place of the dots:

. 9 9 < 2 0 0 ; 3 8 5 > 3 . 5

3 . 5 < 3 0 8 ; 9 1 5 > . 6 5

10 - Write the numbers in decreasing order on the bar:

140 ; 120 ; 250 ; 805 ; 153 ; 270 ; 400

900

11 - Put and calculate:

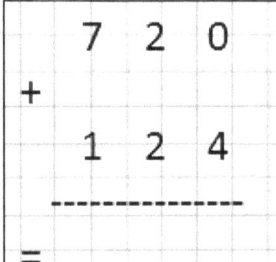

```
    7  2  0
+
    1  2  4
-----------
=
```

```
    5  3  9
+
    3  2  4
-----------
=
```

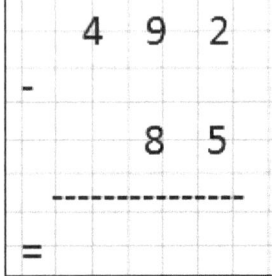

```
    4  9  2
-
       8  5
-----------
=
```

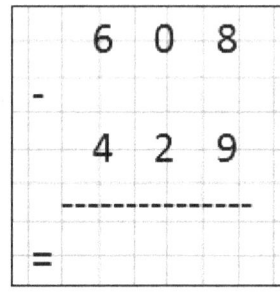

```
    6  0  8
-
    4  2  9
-----------
=
```

12 - Complete the drawing of the two pieces:

5 Centimeters

8 Cm

13 - Anna owns $880, and buys a $420 phone, $24 wallet, and $50 calculator. What is the remaining amount for the girl?

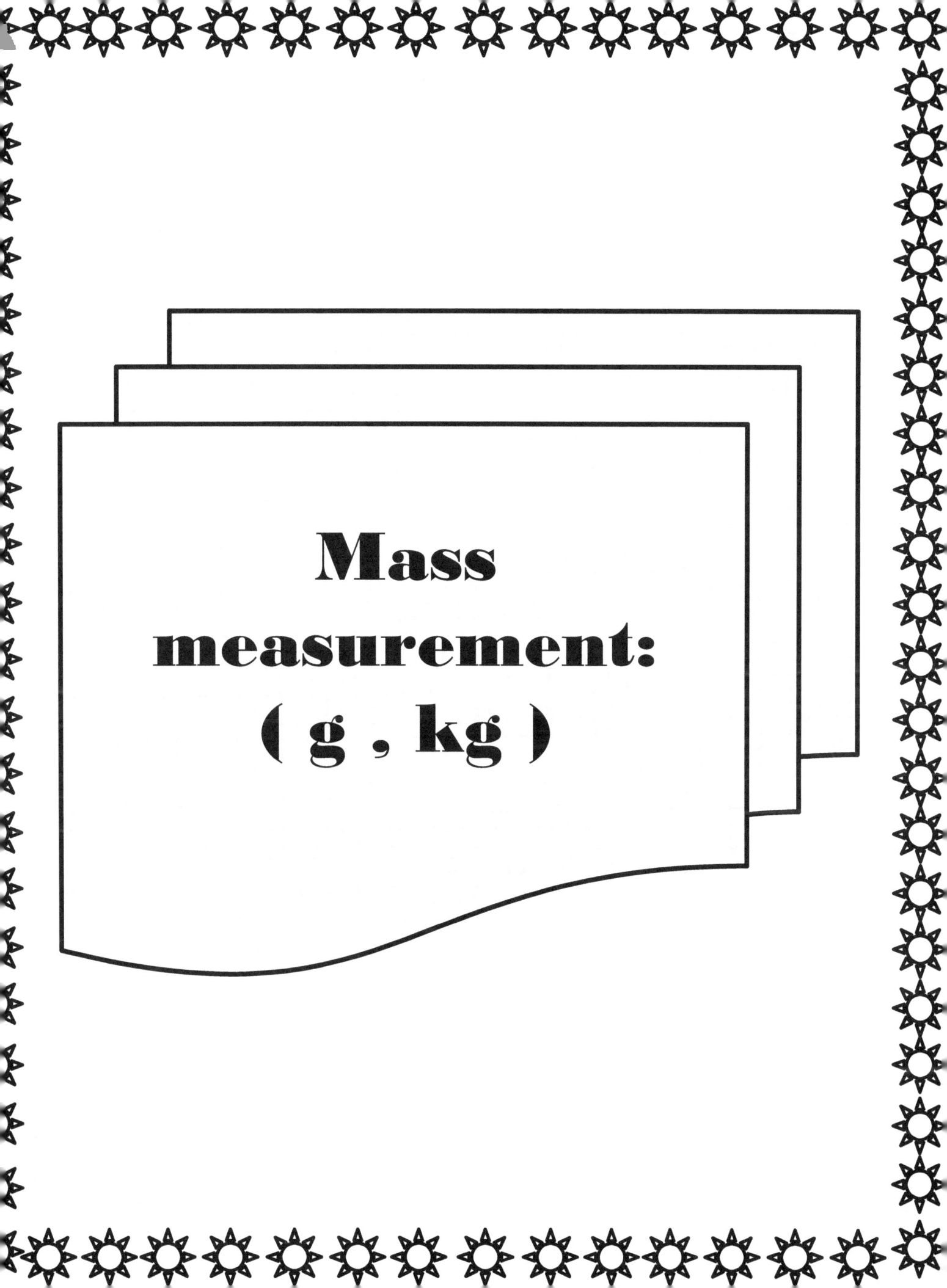

Mass measurement: (g , kg)

Note the scale and write the weight of the fruit:

Apple block:	g

+ =	g

strawberry block :	g

+ =	g

The origin of the right thing with the right mass:

55 g	150 g	450 g

Skip Count by [3] and connect them with Lines from 3 - 99

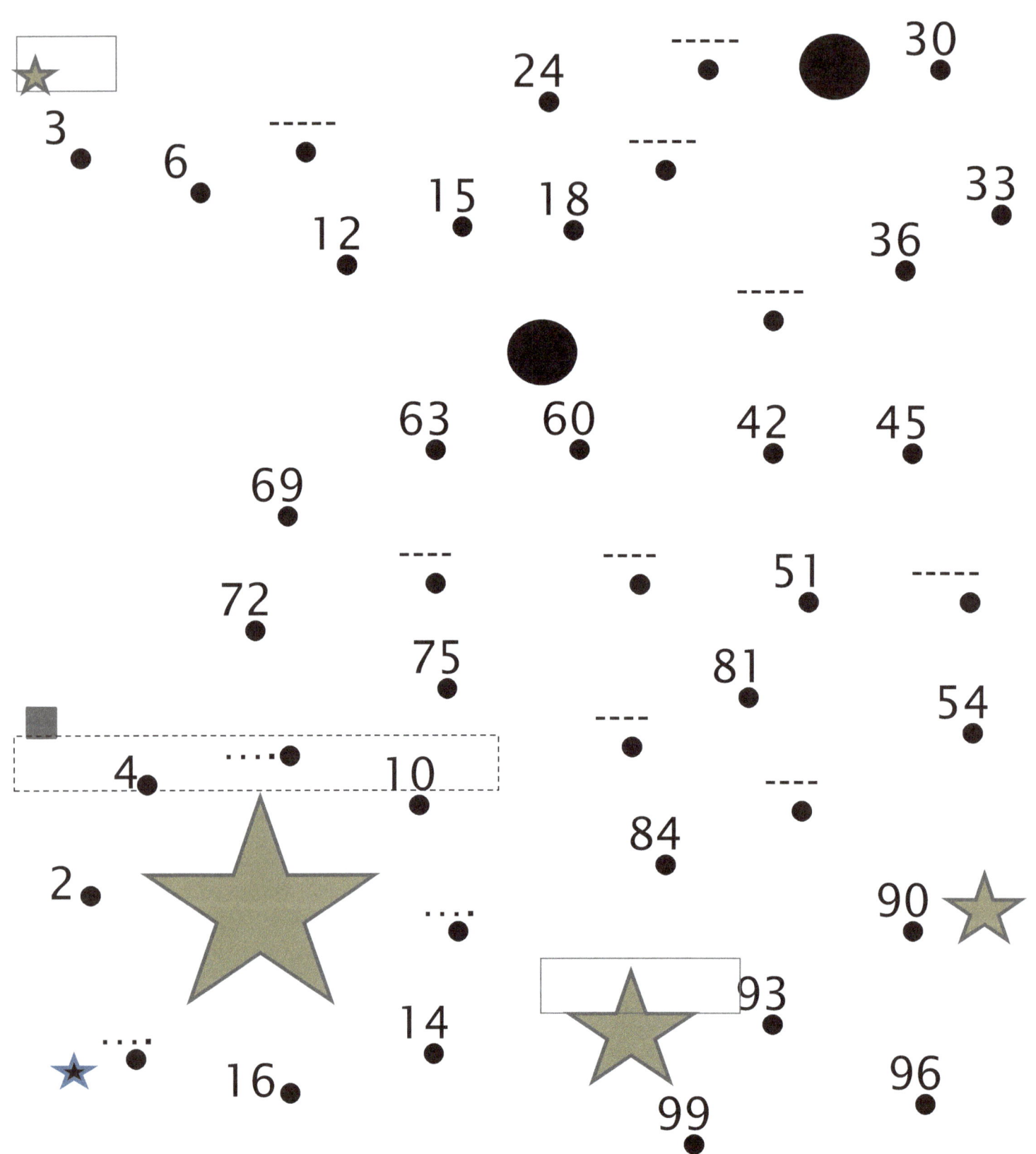

NAME

....

1 2	4 4	1 2	2 6	7 8
+ 3 6	+ 5 4	+ 3 8	+ 5 4	+ 2 1

5 2	2 7	8 7	5 8	3 9
+ 3 5	+ 1 2	+ 0 1	+ 3 1	+ 6 0

0 5	4 4	6 6	5 7	7 7
+ 9 4	+ 5 4	+ 3 1	+ 0 2	+ 1 2

8 + 5 = ____   7 + 2 = ____   11 + 5 = ____

9 + 5 = ____   8 + 3 = ____   14 + 5 = ____

7 + 5 = ____   4 + 9 = ____   15 + 5 = ____

6 + 5 = ____   2 + 6 = ____   19 + 5 = ____

9 5	8 4	6 8	5 6	7 8
+ 3 4	+ 5 4	+ 3 1	+ 5 4	+ 7 7
□ □	□ □	□ □	□ □	□ □

5 2	2 7	8 7	5 8	7 9
+ 3 5	+ 1 2	+ 0 1	+ 3 1	+ 6 0
□ □	□ □	□ □	□ □	□ □

7 5	4 4	6 6	5 7	7 7
+ 1 4	+ 5 4	+ 3 1	+ 0 7	+ 1 2
□ □	□ □	□ □	□ □	□ □

$20 - 3 = \underline{\quad}$ $17 - 6 = \underline{\quad}$ $55 - 15 = \underline{\quad}$

$30 - 8 = \underline{\quad}$ $39 - 5 = \underline{\quad}$ $18 - 5 = \underline{\quad}$

$42 - 5 = \underline{\quad}$ $71 - 20 = \underline{\quad}$ $29 - 19 = \underline{\quad}$

$57 - 9 = \underline{\quad}$ $40 - 5 = \underline{\quad}$ $15 - 15 = \underline{\quad}$

Example

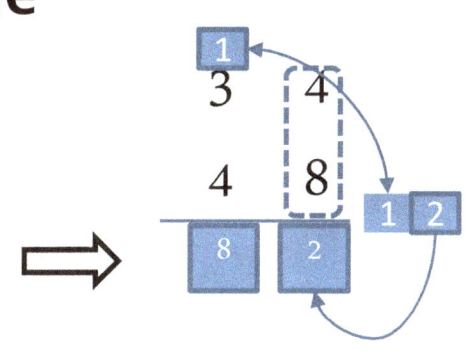

```
  3   5          5   4          6   7          4   6          7   8
+ 4   9        + 3   7        + 4   8        + 5   8        + 4   6
_____      _____      _____      _____      _____


  5   5          2   7          8   7          5   8          3   9
+ 3   9        + 1   5        + 2   6        + 3   4        + 6   8
_____      _____      _____      _____      _____


  4   5          6   6          6   6          5   7          7   7
+ 9   5        + 5   4        + 8   5        + 0   9        + 1   5
_____      _____      _____      _____      _____
```

NAME .... Practice Subtractions

$$35 - 15 =$$

$$15 - 15 =$$

$$25 - 5 =$$

$$48 - 19 =$$

$$12 - 15 =$$

$$28 - 13 =$$

$$8 - 5 =$$

$$42 - 11 =$$

$$50 - 50 =$$

$$25 - 30 =$$

$$25 - 10 =$$

Minus

$$55 - 15 =$$

$$84 - 43 =$$

$$50 - 5 =$$

$$8 - 5 =$$

$$1 - 1 =$$

$$9 - 7 =$$

Additions for fun

$\begin{array}{r} 1\ 2 \\ 1\ 3 \\ \hline \end{array}$ + $\begin{array}{r} 1\ 2 \\ 2\ 0 \\ \hline \end{array}$

$\begin{array}{r} 1\ 0 \\ 2\ 5 \\ \hline \end{array}$ $\begin{array}{r} 1\ 9 \\ 1\ 1 \\ \hline \end{array}$

$\begin{array}{r} 2\ 9 \\ 1\ 7 \\ \hline \end{array}$ =

$\begin{array}{r} 5\ 0 \\ +\ \ \ 0 \\ \hline \end{array}$ =

$\begin{array}{r} +\ 5\ 0 \\ 3\ 0 \\ \hline \end{array}$ = $\begin{array}{r} +\ 9\ 6 \\ 1\ 4 \\ \hline \end{array}$ = $\begin{array}{r} +\ 7\ 5 \\ 2\ 5 \\ \hline \end{array}$ =

8 + 8 =   15 + 35 =

$\begin{array}{r} 3\ 9 \\ 2\ 7 \\ \hline \end{array}$ =

15 + 8 =

25 + 5 =

Name ....

Complete the table below

by adding up the numbers

0	6	12					36
8							
10							70
12		48					

Name ....

Add, subtract and multiply

	+ 3	- 5	+ 5	* 1
42	45	37	47	42
28	-----	-----	-----	-----
52	-----	-----	-----	-----
17	-----	-----	-----	-----
8	-----	-----	-----	-----
44	-----	-----	-----	-----
72	-----	-----	-----	-----
36	-----	-----	-----	-----
9	-----	-----	-----	-----
12	-----	-----	-----	-----
14	-----	-----	-----	-----
32	-----	-----	-----	-----

Name ....

Repeated Additions as Multiplication

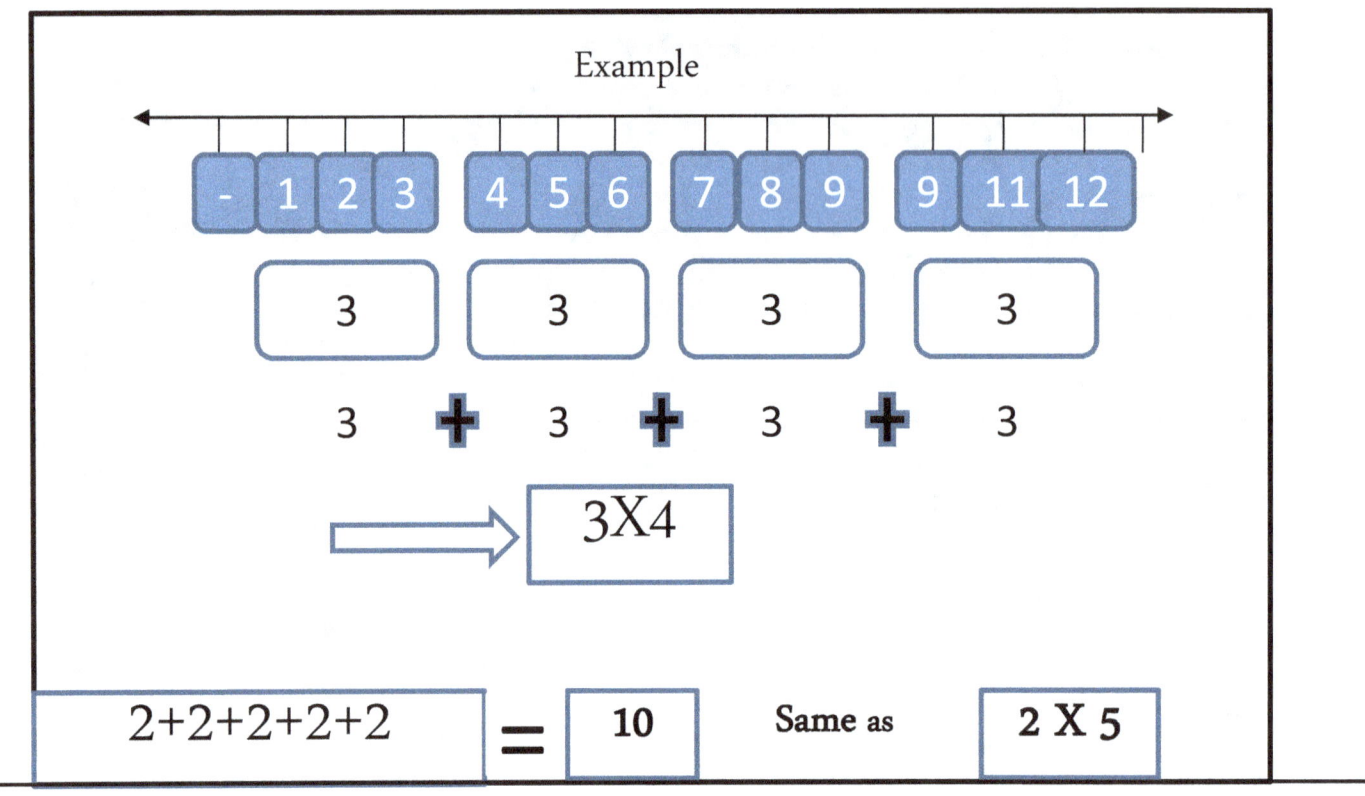

| 2+2+2+2+2 | = | 10 | Same as | 2 X 5 |

Practice Lessons

a | 5+5+5+5+5 | =

____ ☐

c | 6+6+6+6+6 | =

____ ☐

b | 7+7+7+7+7 | =

____ ☐

d | 9+9+9+9+9 | =

____ ☐

NAME...................................
...

What is an array? An array is a set grouped in rows and column

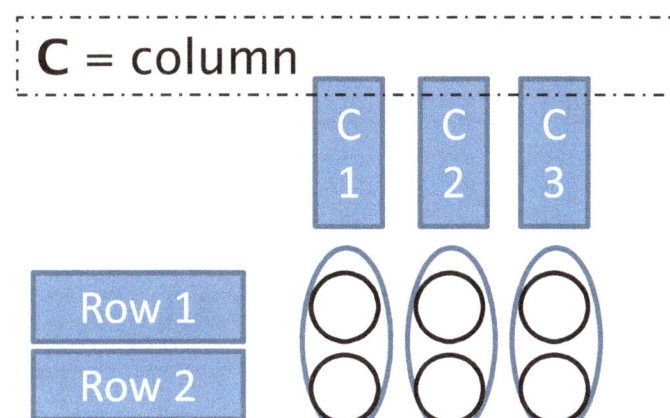

C = column

Row 1
Row 2

Example

You write this as

2 X 3 = 6

(2 rows 3 columns)

Write these as expression equation showing the number of shapes

a

_____

b

_____

c

_____

d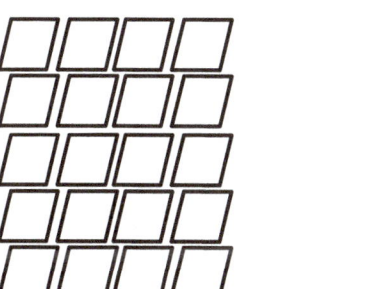

_____

e

_____

f

_____

	4		8			14			20
		26				34	36		
	44				52	54			
	64				72			78	
82			88		92				100
102				110			116	118	
122		126			132			138	
	144			150		154			160
		166					176		180
				190					
202					212				220
			228	230			236		
	244			250	252				260
									280

PRIME NUMBER

NAME ....

A prime number is a number **bigger than one** and that *__cannot__* be divided by any other number *__but itself and one__*.

Examples of prime numbers between **1 - 50**

1	2	3	4	5	6	7	8	9	10
11	12	13	14	15	16	17	18	19	20
21	22	23	24	25	26	27	28	29	30
31	32	33	34	35	36	37	38	39	40
41	42	43	44	45	46	47	48	49	50

These are Prime Numbers

2	3	5	7	11	13	17	19	23
29	31	37	41	43	47			

Even and Odd Numbers

An *even number* can be sorted into pairs and *nothing will remain*, and *an Odd number* always has a remainder.

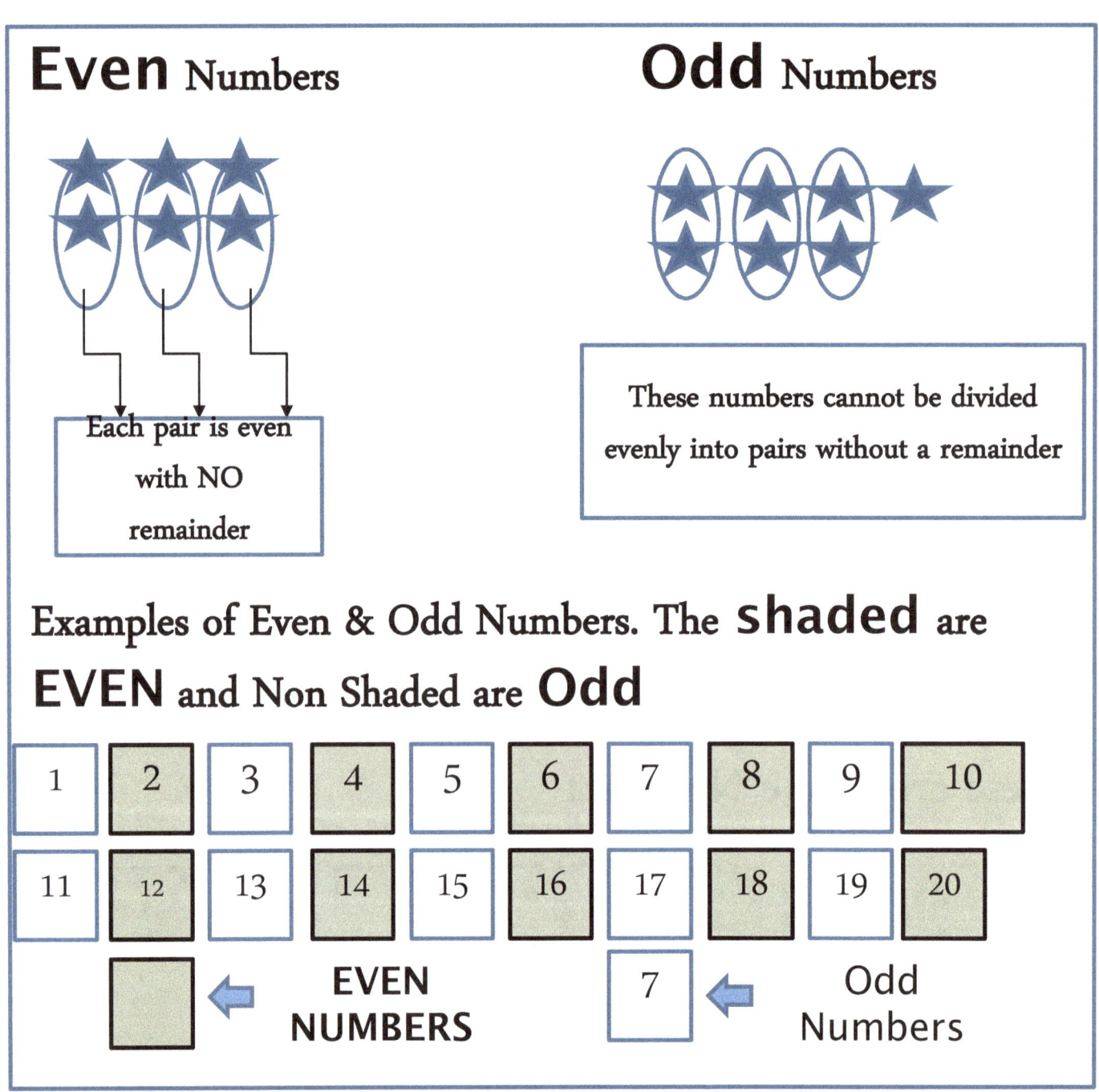

Lets try to write **4 prime**, **even** and **odd** numbers

1	2	3	4	5	6	7	8	9	10
11	12	13	14	15	16	17	18	19	20
21	22	23	24	25	26	27	28	29	30
31	32	33	34	35	36	37	38	39	40
41	42	43	44	45	46	47	48	49	50

	PRIME NUMBERS	EVEN NUMBERS	ODD NUMBERS
a	13	2	3
b			
c			
d			
e			

Select by **circling** all **Even** Numbers amongst the following numbers

3 4 6 7

9 10 8

13 15

11 12 14 24

16 17 23

18 20 22 25

19 21 (2)

Select by circling all **Odd** Numbers in the numbers below

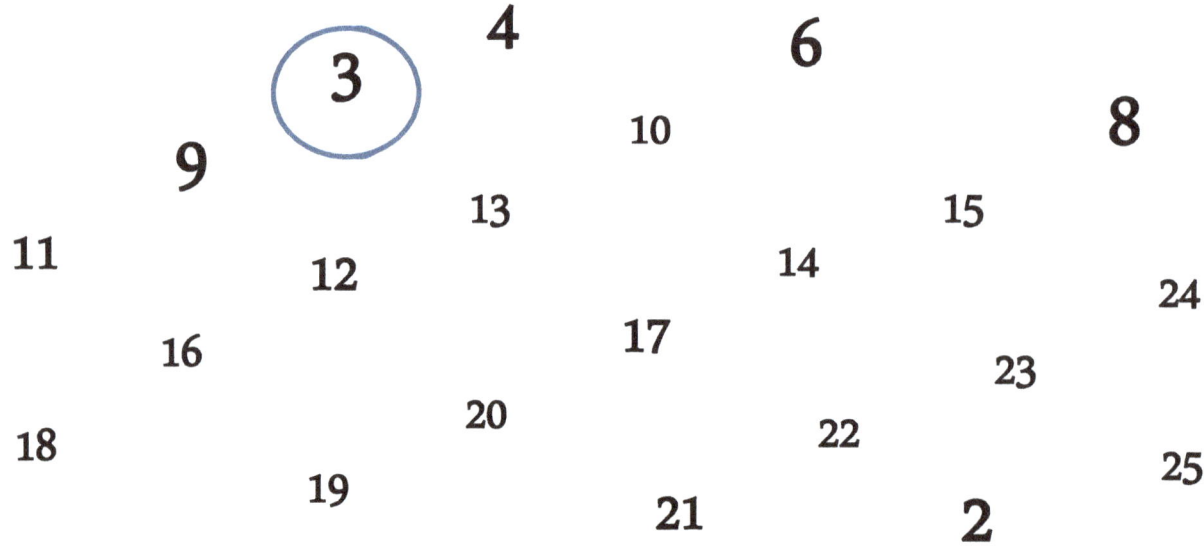

(3) 4 6

9 10 8

13 15

11 12 14 24

16 17 23

18 20 22 25

19 21 2

Show whether each Additions answer is

ODD or Even

1+6 = | 7 | Odd

2+2 = | 4 | Even

3+2 = | |

3+4 = | |

2+7 = | |

3+0 = | |

6+12 = | |

10+3 = | |

Match The following

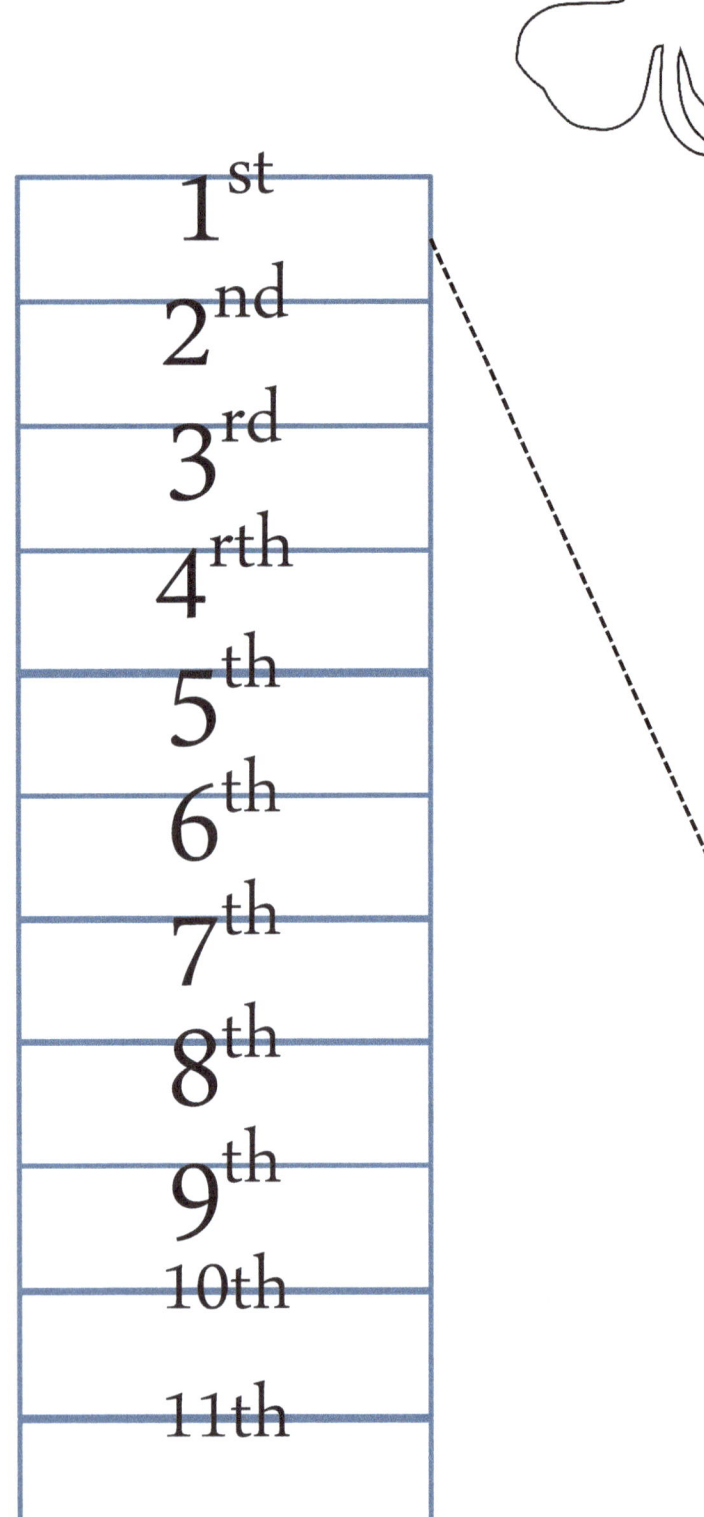

1st

2nd

3rd

4^{rth}

5th

6th

7th

8th

9th

10th

11th

Tenth

Eighth

Six

Second

Ninth

Five

Fourth

Seven

Fifth

first

Third

Eleventh

NAME

....

Example

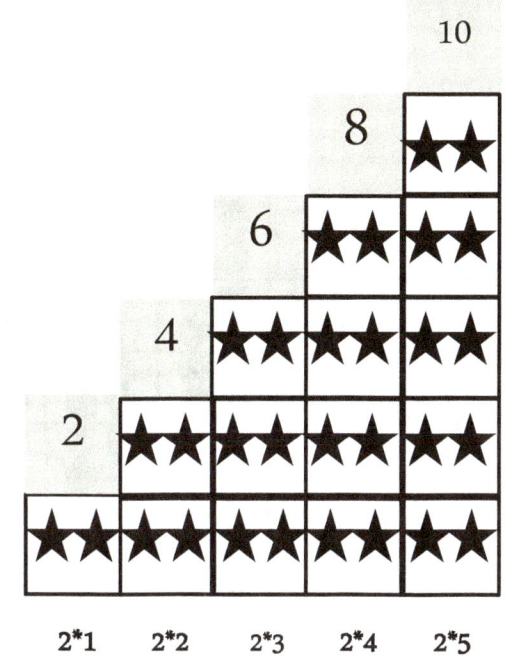

				10
			8	★★
		6	★★★	★★
	4	★★★	★★★	★★
2	★★★	★★★	★★★	★★
★★★	★★★	★★★	★★★	★★

2*1 2*2 2*3 2*4 2*5

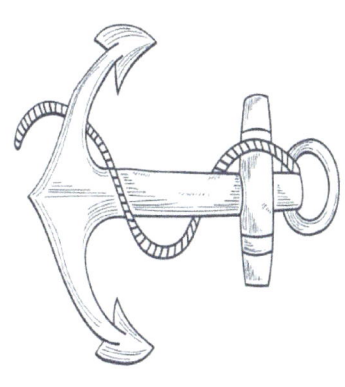

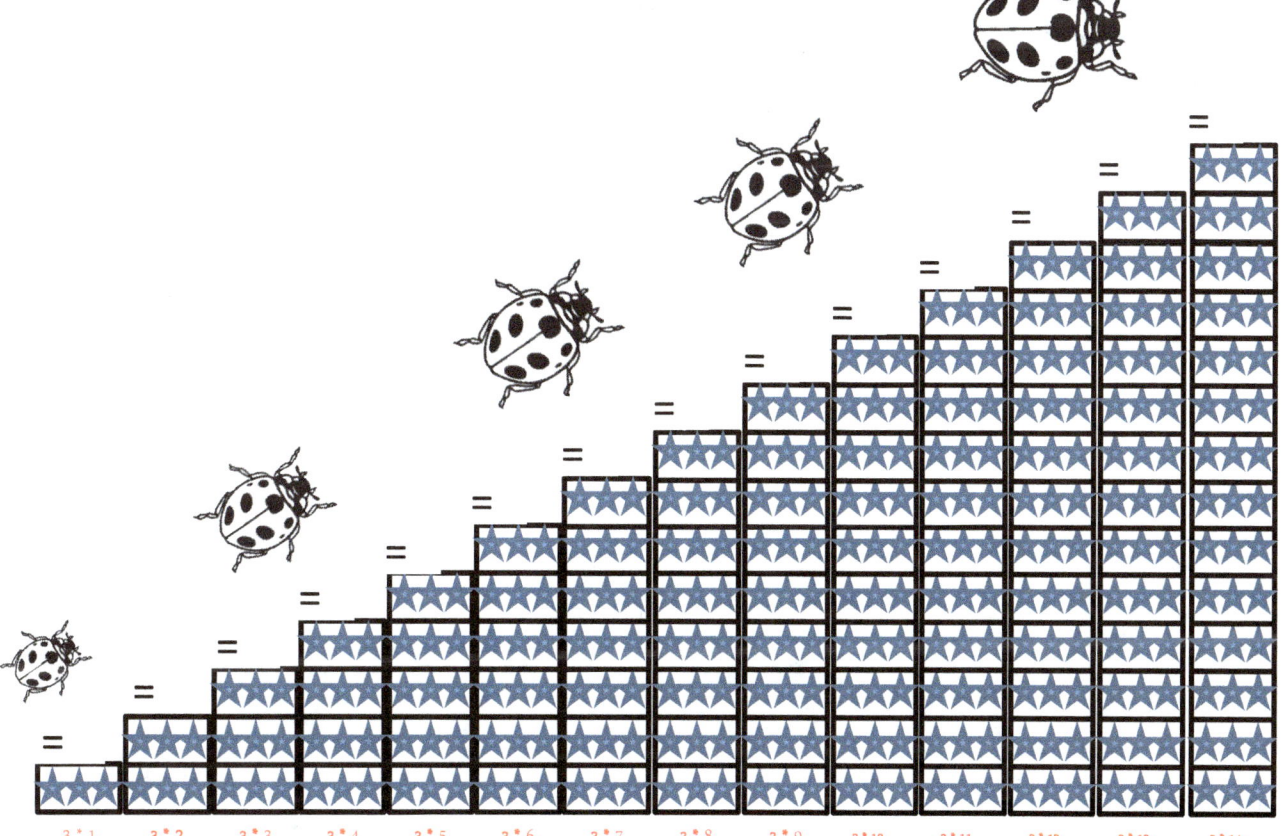

3 * 1 3 * 2 3 * 3 3 * 4 3 * 5 3 * 6 3 * 7 3 * 8 3 * 9 3 * 10 3 * 11 3 * 12 3 * 13 3 * 14

NAME ....

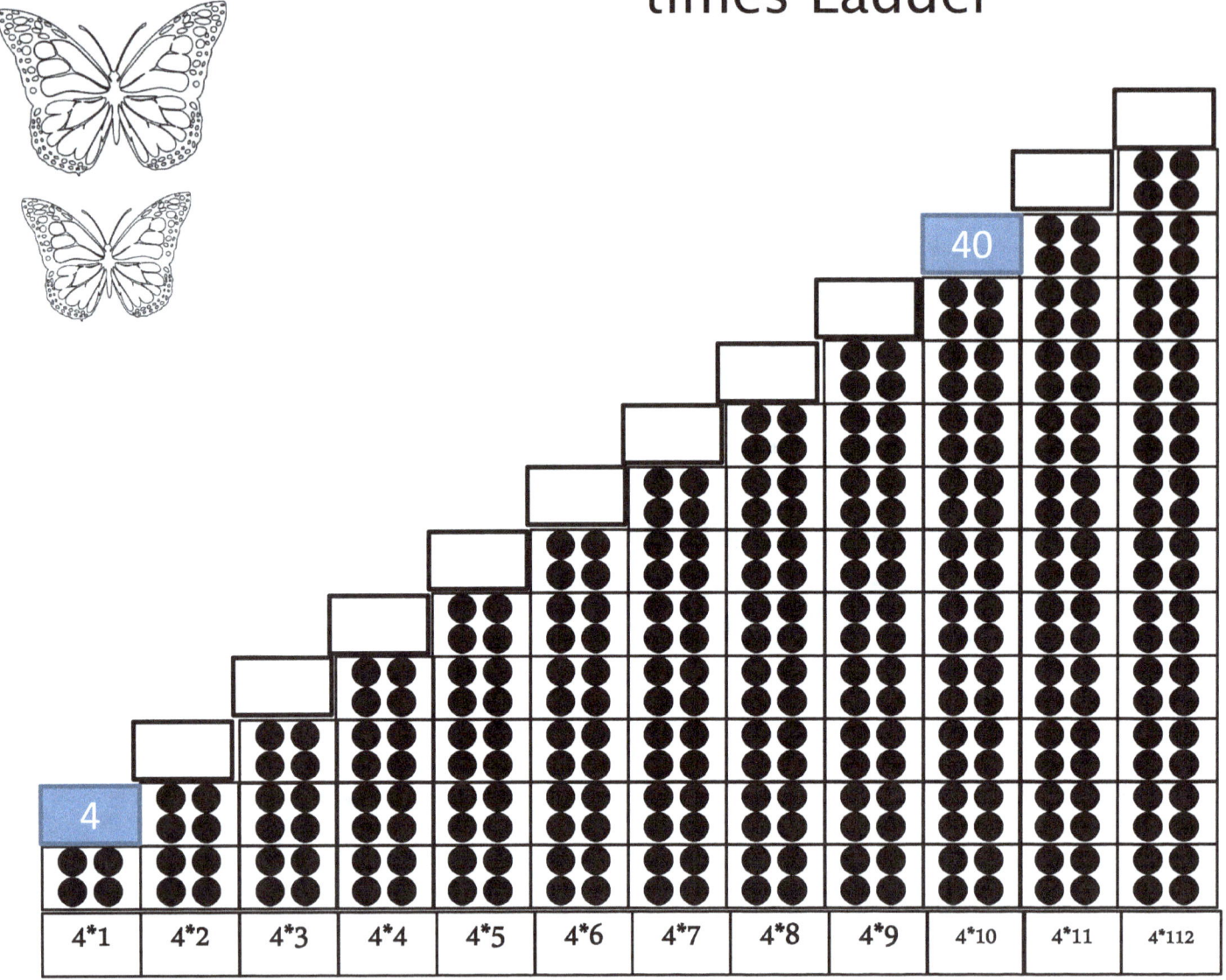

| 4*1 | 4*2 | 4*3 | 4*4 | 4*5 | 4*6 | 4*7 | 4*8 | 4*9 | 4*10 | 4*11 | 4*112 |

4*1 = ...

4*2 = ...

4*3 = ...

4*4 = ...

4*5 = ...

4*6 = ...

4*7 = ...

4*8 = ...

Try your hands on the
following multiplication

5 * 1 ⮕ = _____
5 * 2 ⮕ = _____
5 * 3 ⮕ = _____
5 * 4 ⮕ = _____
5 * 5 ⮕ = _____
5 * 6 ⮕ = _____
5 * 7 ⮕ = _____
5 * 8 ⮕ = _____
5 * 9 ⮕ = _____
5 * 10 ⮕ = _____
5 * 11 ⮕ = _____
5 * 12 ⮕ = _____
5 * 13 ⮕ = _____
5 * 14 ⮕ = _____
5 * 15 ⮕ = _____
5 * 16 ⮕ = _____
5 * 17 ⮕ = _____
5 * 17 ⮕ = _____

6 * 1 ⮕ = _____
6 * 2 ⮕ = _____
6 * 3 ⮕ = _____
6 * 4 ⮕ = _____
6 * 5 ⮕ = _____
6 * 6 ⮕ = _____
6 * 7 ⮕ = _____
6 * 8 ⮕ = _____
6 * 9 ⮕ = _____
6 * 10 ⮕ = _____
7 * 1 ⮕ = _____
7 * 2 ⮕ = _____
7 * 3 ⮕ = _____
7 * 4 ⮕ = _____
7 * 5 ⮕ = _____
7 * 6 ⮕ = _____
7 * 7 ⮕ = _____
7 * 8 ⮕ = _____

Multiplication

Example

2	*	1	→	2		=	**2**	3	*	1	→	3		=	**3**

2 * 1 → 2 = **2**   3 * 1 → 3 = **3**

2 * 2 → 2+2 = **4**   3 * 2 → 3+3 = **6**

2 * 3 → 2+2+2 = **6**   3 * 3 → 3+3+3 = **9**

2 * 4 → 2+2+2+2 = **8**   3 * 4 → 3+3+3+3 = **12**

2 * 5 → 2+2+2+2+2 = **10** 3 * 5 → 3+3+3+3+3 = **15**

---

3 * 1 → = ____   4 * 1 → = ____

3 * 2 → = ____   4 * 2 → = ____

3 * 3 → = ____   4 * 3 → = ____

3 * 4 → = ____   4 * 4 → = ____

3 * 5 → = ____   4 * 5 → = ____

3 * 6 → = ____   4 * 6 → = ____

3 * 7 → = ____   4 * 7 → = ____

3 * 8 → = ____   4 * 8 → = ____

3 * 9 → = ____   4 * 9 → = ____

3 * 10 → = ____   4 * 10 → = ____

Multiplication Boxes

Examples

a

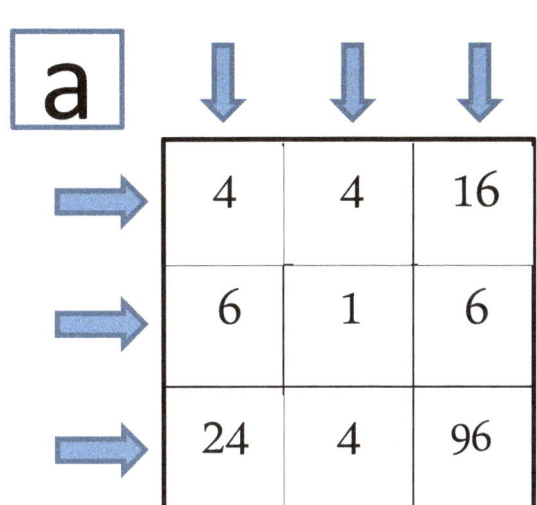

4	4	16
6	1	6
24	4	96

b

6		42
6		18
		756

c

⁻3	3	
	3	18
-18		

c

	7	49
35	7	

d

1		⁻10
⁻70		
		70

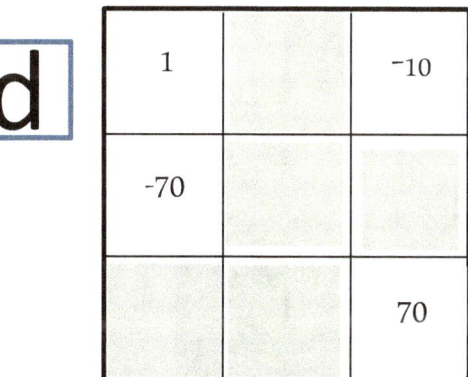

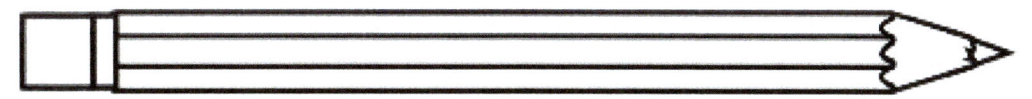

Fractions

Examples of Fractions

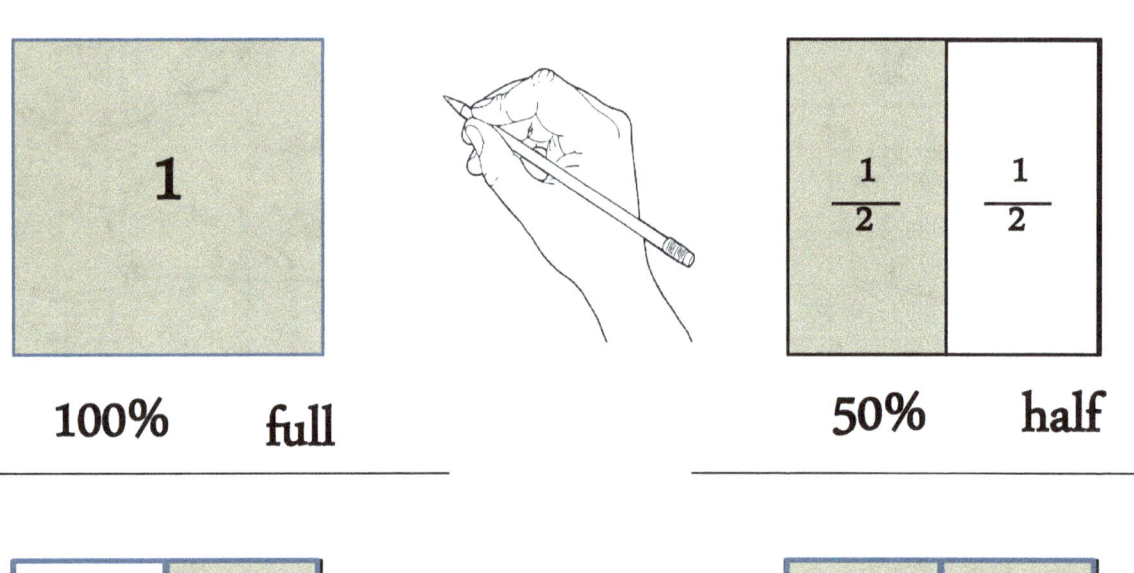

100%   full

50%   half

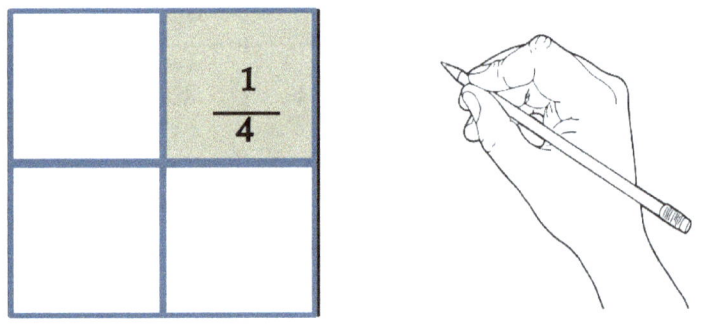

25%   Quarter

3 Quarters

75%

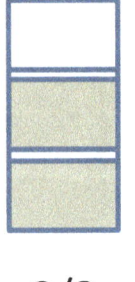

2/3

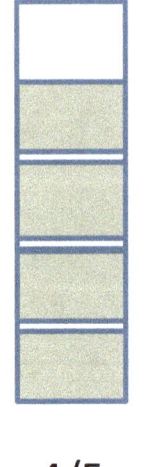

4/5

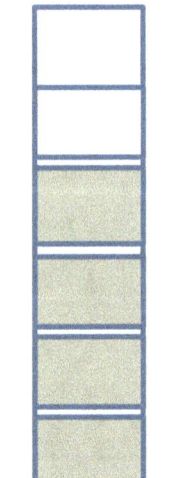

4/6

3/7

Fractions

Choose the right answers

a

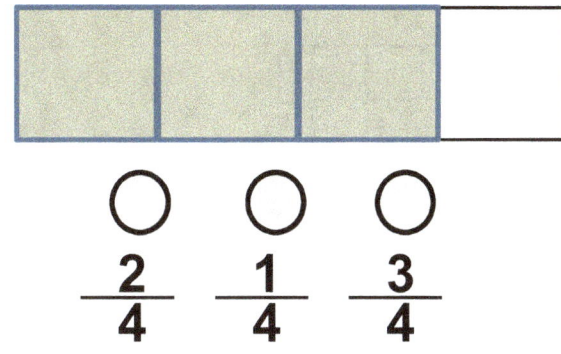

$\frac{2}{4}$ $\frac{1}{4}$ $\frac{3}{4}$

b

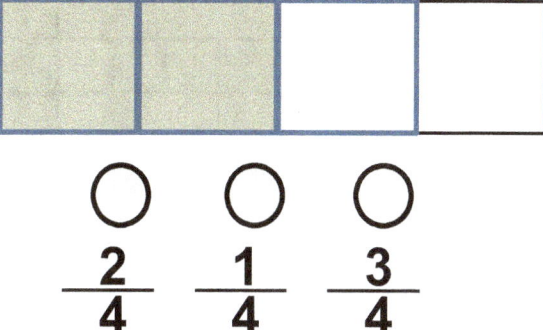

$\frac{2}{4}$ $\frac{1}{4}$ $\frac{3}{4}$

c

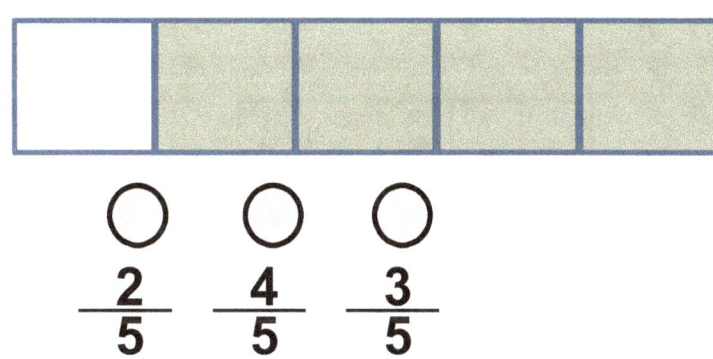

$\frac{2}{5}$ $\frac{4}{5}$ $\frac{3}{5}$

What % is this

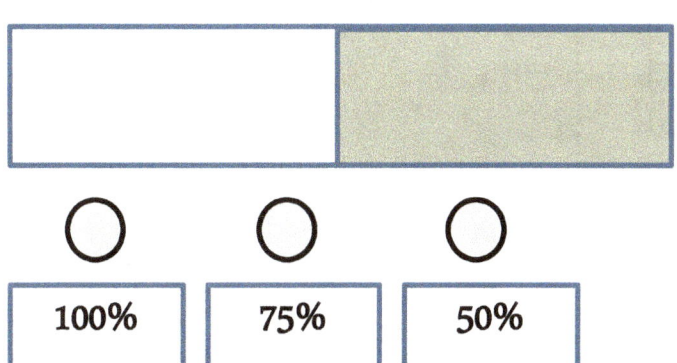

100%   75%   50%

d

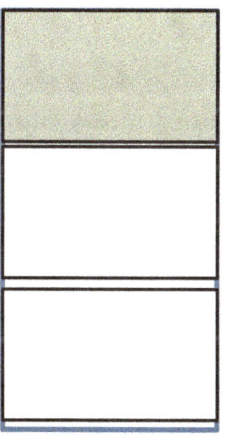

$\frac{1}{3}$ $\frac{2}{3}$ $\frac{3}{3}$

Examples

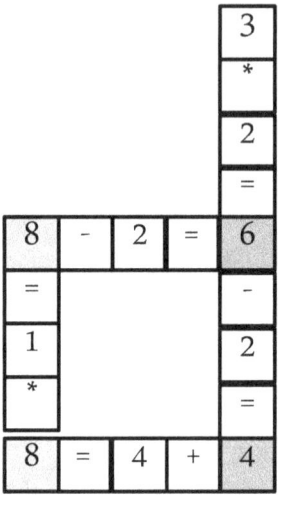

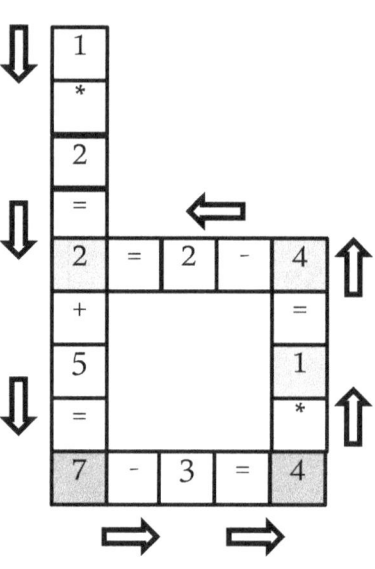

Try this

Name ....

More Examples to be tried

			2			
3	*	3	+	2	=	
			=			
			12	*	1	=		+	12
			+						=
									24

	7	=	7	-	10
30	-		=	30		*
-						
27						=
*		-	28	=	20	+		=	32
4
+
3
=

NAME

....

Take your cool, and solve the

following **maths** challenge

Examples

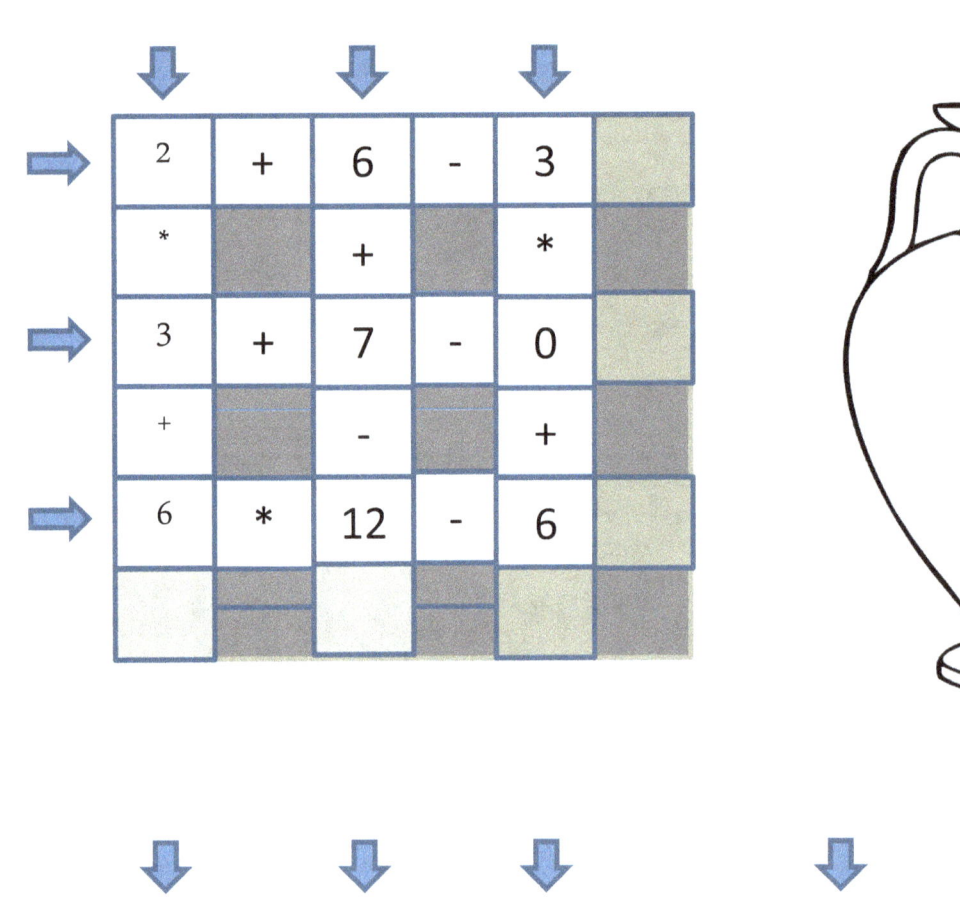

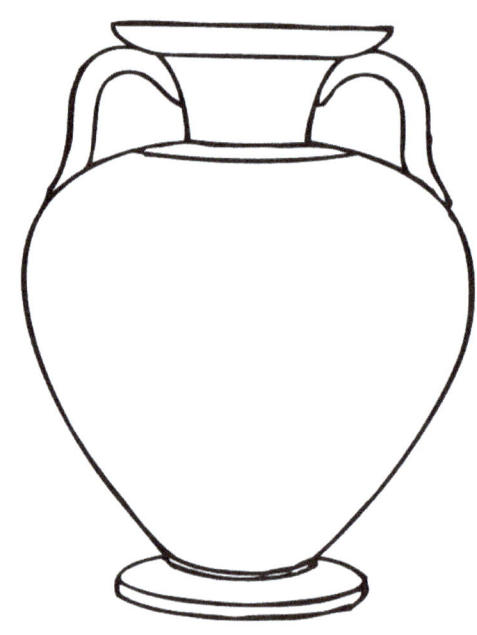

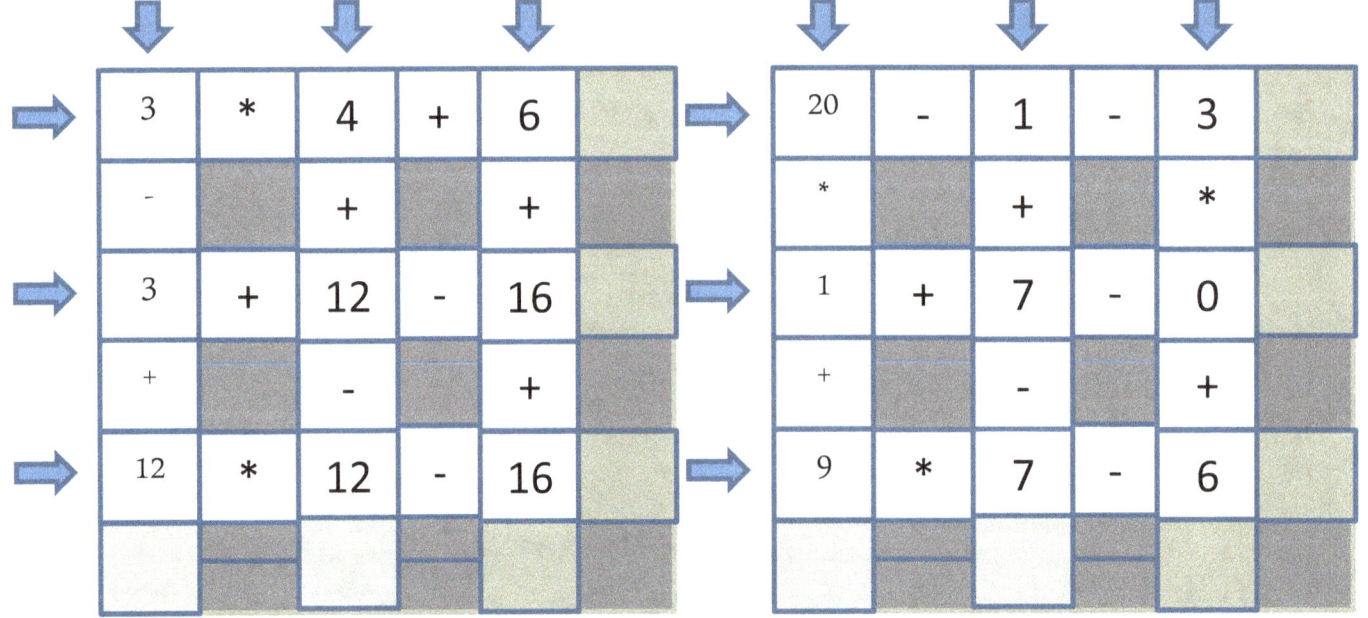

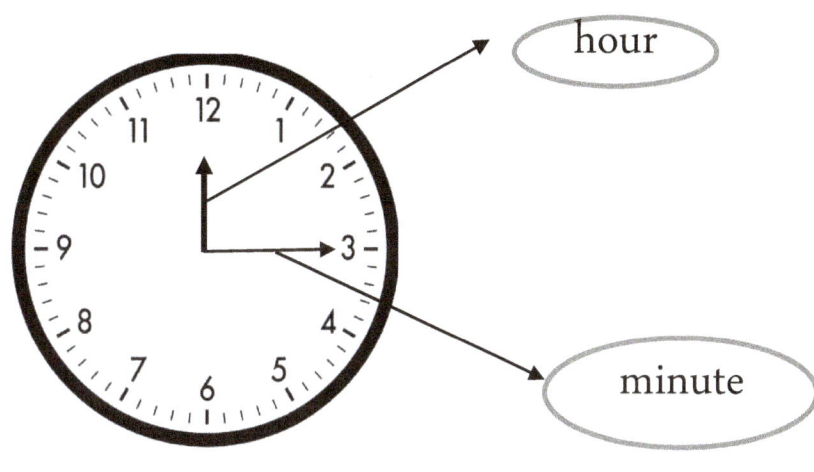

hour

minute

12hr:15min

60 Seconds	=	1 minute

60 minutes	=	1 hour

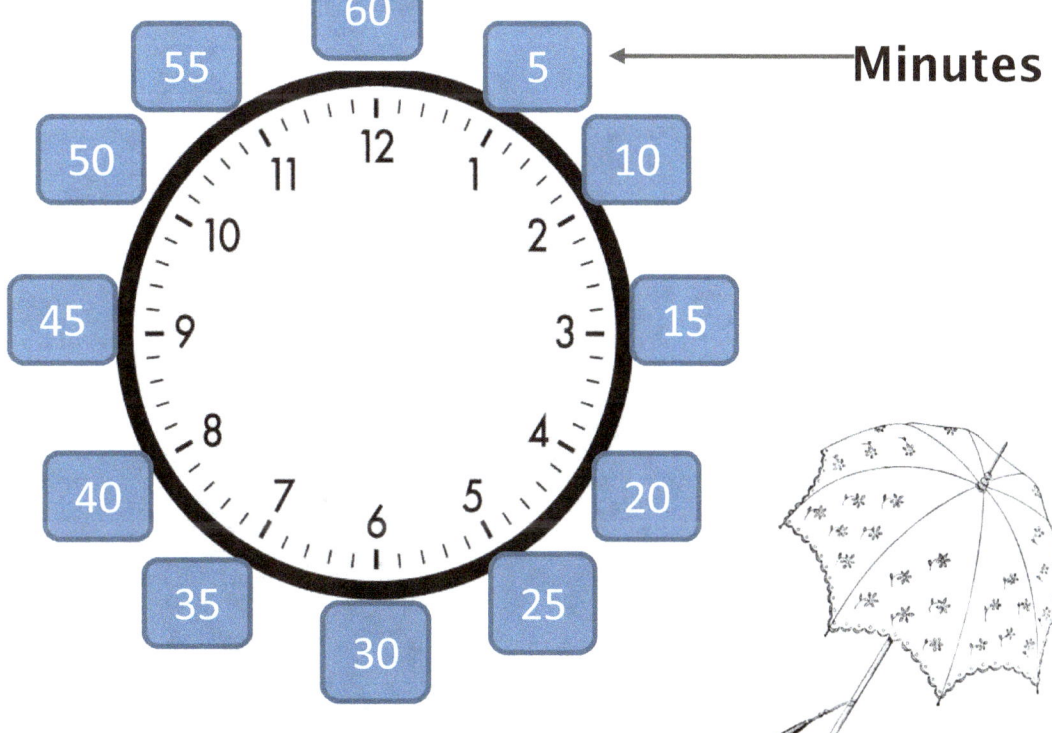

Minutes

NAME ....

Convert the **minutes** into

hours

a   180 min = _____ hr   e   300 min = _____ hr

b   120 min = _____ hr   d   360 min = _____ hr

c   240 min = _____ hr   e   130 min = _____ hr

d   60 min = _____ hr   f   140 min = _____ hr

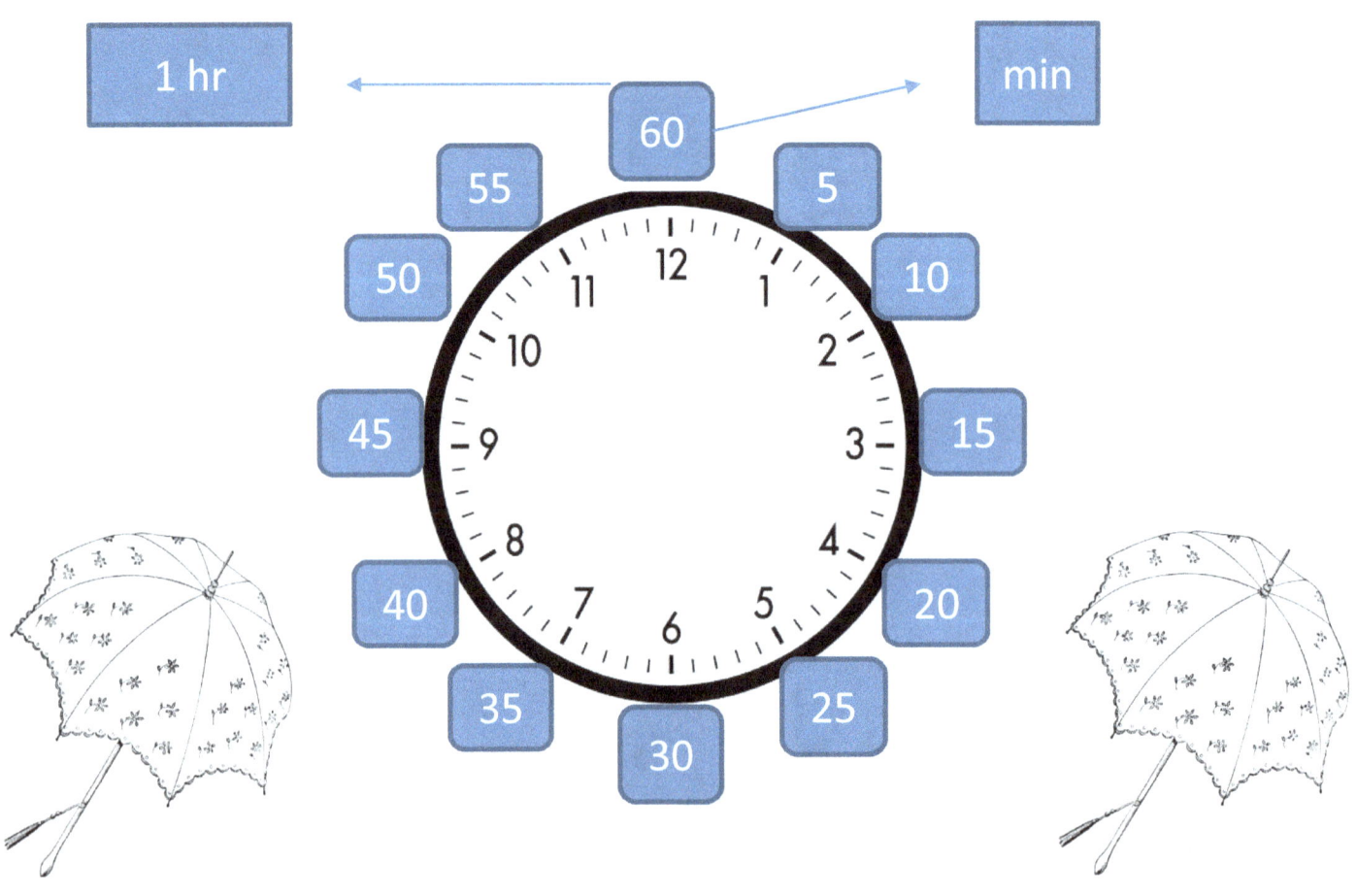

Examples

12:00

12:15

12:45

12:30

Write the correct time

NAME..........................

Plot the **time** on the **clock**

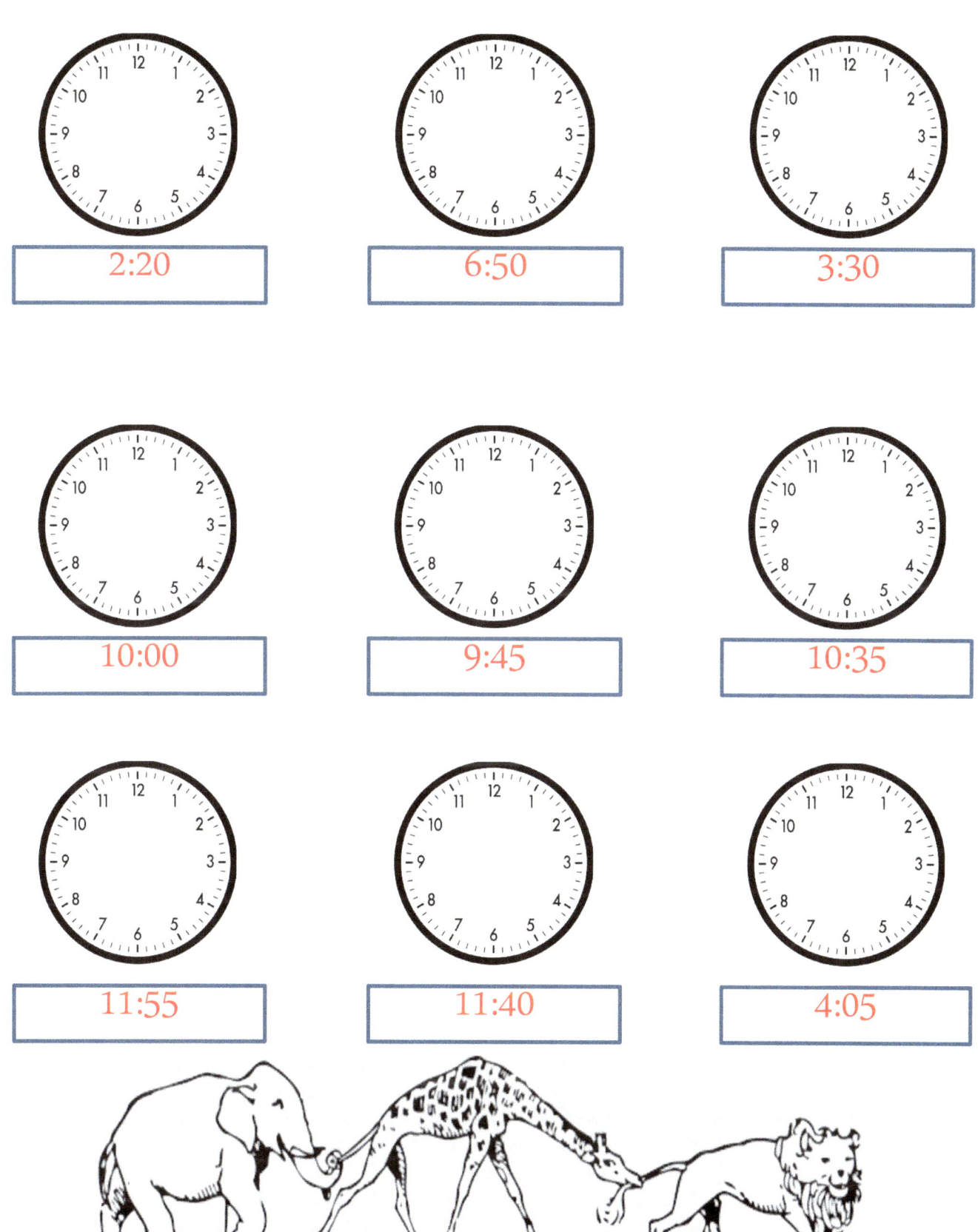

2:20

6:50

3:30

10:00

9:45

10:35

11:55

11:40

4:05

NAME .... # Months and Days

	MONTH	NO OF DAYS
1	January	31
2	February	28
3	March	31
4	April	30
5	May	31
6	June	30
7	July	30
8	August	30
9	September	30
10	October	31
11	November	31
12	December	31

Fill the forms with the
Number of days and name of
the months

	MONTH	NO OF DAYS
1	January	31
2	February	28
3	....	....
4	April	30
....	May	31
6	June	....
7	....	30
8	August	30
9	September	....
10	October	31
	....	
11	...	....
12	December	31

www.ingramcontent.com/pod-product-compliance
Lightning Source LLC
Chambersburg PA
CBHW081451220526
45466CB00008B/2588